CHAOTIC ECONOMIC DYNAMICS

Chaotic Economic Dynamics

RICHARD M. GOODWIN

CLARENDON PRESS OXFORD
1990

Oxford University Press, Walton Street, Oxford OX2 6DP.

Oxford New York Toronto
Delhi Bombay Calcutta Madras Karachi
Petaling Jaya Singapore Hong Kong Tokyo
Nairobi Dar es Salaam Cape Town
Melbourne Auckland

and associated companies in
Berlin Ibadan

Oxford is a trade mark of Oxford University Press

Published in the United States
by Oxford University Press, New York

British Library Cataloguing in Publication Data
Goodwin, R. M. (Richard Murphey) 1913–
Chaotic economic dynamics.
1. Economics. Theories
I. Title
330.1
ISBN 0–19–828335–0

Library of Congress Cataloging in Publication Data
Goodwin, Richard M. (Richard Murphey), 1913–
Chaotic economic dynamics/Richard M. Goodwin.
Includes bibliographical references.
1. Statics and dynamics (Social sciences)   2. Economics,
Mathematical.   3. Chaotic behavior in systems.   I. Title.
HB145.G66   1990   330'.01'51—dc20   90–32566
ISBN 0–19–828335–0

Typeset by Butler & Tanner Ltd, Frome and London
Printed in Great Britain by
Bookcraft Ltd
Midsomer Norton, Avon

Preface

THE origin of this collection of short essays was a series of seminars given in 1988 at the European University Institute in Florence, Italy. My aim has been to elaborate the central conceptual framework of the modern industrial economy. In this sense it derives from the formulation of the problem by my teacher and friend Joseph Schumpeter. Though a neoclassical economist, he perceived the essentially evolutionary nature of the industrialized nations.

By comparison with the natural sciences, economics suffers from the lack of a solid empirical foundation based on generally valid experimental data. To make up for this deficiency, an ingenious substitute has been elaborated with great subtlety and considerable success. The method consists in asking what would a rational man (now fashionably called an 'agent') do when confronted by the manifold problems of an economic nature: he is alleged to maximize his utility or his satisfactions, by minimizing his costs and maximizing his profits or his gains of whatever sort he desires. Under the banner of General Equilibrium Theory, this has been developed into an imposing analytic web of how a system of a large number of such agents would interact in a unified market mechanism. This programme, in an increasingly mathematical form, has produced impressive results, which may be considered 'mainstream' economics. Some tentative efforts at a kind of experimental economics have raised serious doubts about this 'rational' behaviour.

This analysis appeared to work well for a single moment in time, but there always remained the awkward fact that both agents and goods have a future. So while one could in principle solve simultaneously for all prices and quantities at a point of time, one really needed the impossibly difficult set of solutions also over the infinite future! Since the future had to be regarded as unknown and hence uncertain, it all needed to be reformulated as a gambling game. John von Neumann's formulation of game theory proved too weak a tool to resolve such a gigantic problem.

It is at this point that economic dynamics becomes relevant. It has been proposed, quite naturally but surely unfortunately, to deploy the powerful tool of rationality to decisions in dynamics. One can apply rational choice to known prices and quantities but not to unknown future ones. The subject of these essays is chaotic dynamics, the most arresting consequence of which is that a completely deterministic system produces unpredictable behaviour—unpredictable in the sense that looking at the past and the present one cannot say precisely what the future will be. Therefore the very basis empirically of rational prediction is destroyed. Yet agents must and do take decisions, and the aim here is to incorporate such evidence as is easily available as to their behaviour into deterministic systems, the nature of which can be analysed and solved for their behaviour over time.

The character of the model developed here is qualitative rather than realistically quantitative. Only if one has a fruitful analytic scheme can one arrive at satisfactory quantitative results. As I see it one needs a system capable of endogenous, irregular, wavelike growth. The recent discovery and elaboration of 'chaotic attractors' seemed to me to provide the kind of conceptualization that we economists need. Therefore I have tried to provide examples of a number of different simple dynamic models, including both difference and differential varieties. Difference equation systems, though of limited applicability to economics, are included because they provide the simplest, reasonably complete introduction to chaotic analysis: in particular, they require no more than one dimension, by contrast with differential equations which require at least three dimensions.

The grave shortcoming of these essays is that they are aggregative. Schumpeter rightly insisted that innovative technology was essentially specific to particular industries or sectors of the economy. However, satisfactory dynamic analysis of such multisectoral systems is a large and difficult task, requiring the kind of quantitative logic deployed by John von Neumann, and well beyond the scope of this short collection of essays. Unsatisfactory as they are, my hope is that these essays do illuminate the nature of some of the problems, even if not the nature of economic reality.

R.M.G.

Peterhouse, Cambridge
and
University of Siena
1989

Contents

1

Capitalism as Creative, Chaotic Evolution by Structural Change

WHY is economics like the weather?: because both are highly irregular if not chaotic, thus making prediction unreliable or even impossible. Lorenz, a pupil of the great George Birkhoff, opened up one of the more active parts of current mathematical analysis by proposing an endogenous meteorological theory which, without external disturbance, could produce chaotic time-series solutions. This suggests to me that there will have to be a thorough reformulation of much econometrics. It opens up a wide range of problems and possibilities for the use of such models in empirical economics. There are two possible ways of explaining the complexities of economic time-series— exogenous shocks and endogenous models which can produce irregular solutions. There is no question that in analysing economic time-series one must assume irregular shocks, but surely also one must seek out plausible non-linear models to test, models capable of independently producing erratic behaviour, and then separate out the two components. I propose to give a brief, somewhat superficial, account of such non-linear models followed by illustrative examples.

There are two broad types of dynamical equation systems: continuous time and discrete time; both have frequently been used in economics. The latter arise because there are significant time-lags in an economy. The trouble is that these occur in the context of economic activity which is substantially continuous, so that one should formulate mixed difference-differential systems, a procedure the complications of which place it beyond the scope of this book.

Although most economic time-series arrive in discrete time,

1

this is irrelevant since the reality is continuous: this implies differential equations. Simple one- and two-dimensional systems only give erratic behaviour if they have exogenous, time-dependent parameters, for example economies subject to balance-of-payments effects. It appears that for differential systems endogenous chaos can only occur in models of three or more dimensions.

The weird and wonderful world that non-linear dynamics opens up is so complicated that it is highly desirable to keep to low-dimensional models, which I shall do. Yet there are a very large number of dimensions to all economies. Many dimensions usually mean very complicated behaviour, which makes the testing for structure difficult or impossible. Thus even with many, well-defined, constituent cycles, if some of the periods are incommensurate, the solutions will *never* repeat.

To indicate the complicated nature of a multidimensional system embodying structural change, I propose to give a brief example with both price and quantity changes in first one and then a second change in structural parameters. The maximum number of dimensions which my computer will handle is four, so that I shall consider a two-sector economy.

I start with an economic potential defined as

$$V(p,\ q) = \langle p_1,\ p_2 \rangle \left\{ \begin{matrix} q_1 \\ q_2 \end{matrix} \right\} - \langle p_1,\ p_2;\ w \rangle \begin{bmatrix} a_{11} & a_{12} & a_{c1} \\ a_{21} & a_{22} & a_{c2} \\ a_{11} & a_{12} & a_{11} \end{bmatrix}$$

$$\left\{ \begin{matrix} q_1 \\ q_2 \end{matrix} \right\} - \langle B_1,\ B_2 \rangle \left\{ \begin{matrix} q_1 \\ q_2 \end{matrix} \right\} - \langle p_1,\ p_2 \rangle \left\{ \begin{matrix} A_1 \\ A_2 \end{matrix} \right\},$$

where B are constant nominal costs per unit of output and A all real demands not dependent on output. For simplicity I set all $= 0$.

First it is essential to determine the fixed, or equilibrium, points:

$$\hat{p} = \langle w \langle a_1 \rangle + \langle B \rangle \rangle [I - a]^{-1}$$
$$\hat{q} = [I - a]^{-1} \{ \{ a_c \} 1 + \{ A \} \}.$$

Changes in B and in A are considered as exogenous to this simple linear production system. Our principal interest for the moment is the effect of changes in a_{ij}, in a_1, and in a_c. These changes may or may not be exogenous, but in any case they change the character of the dynamical behaviour of the system in response to changes in B and in A.

To find the dynamics, we take the slopes of the potential first with respect to p and then with respect to q. Taking account of consumption $\hat{a}_c$, we have

$$\text{Gradient } V_p = [I - a]\{q\} - l\{\hat{a}_c\} - \{A\}.$$

The simplest dynamical adjustment is that output slides down the hills of demand-output, thus

$$\dot{q} = - \text{Grad. } V_p \text{ and down the hills of value thus}$$
$$\dot{p} = - \text{Grad. } V_q.$$

To consider technical progress and the problem of whether or not it leads to technological unemployment, I shall take a somewhat Schumpeterian view, without the implicit assumption of full employment. In the above potential, there is the assumption that consumption occurs in fixed amounts per employed person (and that no profits are consumed). This assumption is quite unjustified for consumption, unlike production. Therefore, for this problem I take a constant nominal wage rate and assume that all personal income is spent on consumption in a fixed proportion, thus:

$$w = s_1 w + (1 - s_1)w \text{ (ignoring personal services and setting}$$
$$w = 1).$$
$$s_2 = 1 - s_1.$$

Real consumption then becomes inversely related to prices, thus:

$$\text{cons}_1 + \text{cons}_2 = ls_1/p_1 + l(1 - s_1)/p_2.$$

We can then formulate our system of four differential equations:

$$\dot{q}_1 = (a_{11}q_1 + a_{12}q_2 + (a_{11}q_1 + a_{12}q_2)s_1/p_1 + a_1 - q_1)\varepsilon_1$$
$$\dot{q}_2 = (a_{22}q_1 + a_{22}q_2 + (a_{11}q_1 + a_{22}q_2)(1 - s_1)/p_2 + a_2 - q_2)\varepsilon_2$$
$$\dot{p}_1 = (a_{11}p_1 + a_{22}p_2 + a_{11} + b_1 - p_1)\gamma_1$$
$$\dot{p}_2 = (a_{12}p_1 + a_{22}p_2 + a_{22} + b_2 - p_2)\gamma_2.$$

I propose to use this set of equations to analyse the nature of economic growth: to do this I shall utilize the complementary theories of Schumpeter and Keynes. It was originally Schumpeter who posed the problem in its most basic form: one of his formulations states that a

sequence of phenomena leads up to a new neighbourhood of equilibrium, in which enterprise will start again. This new neighbourhood of equilibrium is characterized, as compared to the one that preceded it, by a 'greater' social product of a different pattern, new production functions, equal sum total of money incomes, a minimum (strictly zero) rate of interest, zero profits, zero loans, a different system of prices and a lower level of prices, the fundamental expression of the fact that all the lasting achievements of the particular spurt of innovation have been handed to consumers in the shape of increased real incomes (*Business Cycles*, p. 137).

This would have said it all, had he only written 'greater *potential* social product'. Implicit is the assumption of full employment: but how could he do so in the middle of the great depression of the 1930s and after Keynes? Logically, but unfortunately, he rejected totally the *General Theory*. One should note that in his description of the new equilibrium he includes zero profits, zero innovational investment; had he included continuing high investment, his conclusion would be more acceptable. The system of equations above assumes that output is adjusted to approach demand. Technological innovation means necessarily lowered inputs per unit of output, so that for any level of output demand is reduced, leading to reduction of demand and hence of output. Consequently the simple dynamics mean a falling to a lowered output, and this is the first case I shall consider but not the only one.

In agreement with Schumpeter, the money wage, w, is constant and may be conveniently set equal to unity. All earned income is spent for consumer goods in the fixed proportions, s_1 and $1 - s_1$. Eschewing more sophisticated and complicated analysis, I take real consumption per employee $= s_1/p_1 + (1 - s_1)/p_2$, employment is $l = a_{11}q_1 + a_{12}q_2$. In order merely to indicate the nature of the problem, I assign somewhat arbitrary values to all parameters, leaving the four variables to be determined.

With all variables initially in equilibrium assume that a dramatic innovation permits a reduction of a_{12} from 0.20 to 0.10. The resulting trajectories are shown in Fig. 1.1 for 70 periods with the following percentage changes:

$q_1 - 22.6\%$
$q_2 - 14.5\%$
$p_1 - 4.4\%$
$p_2 - 19.5\%$.

Though the time required to complete the absorption of the innovation differs, it is clear that the effects are completed well before 70. Total employment fell from 1.36 to 0.67, i.e. -49.0%. This result shows how wrong Schumpeter was: though these particular numbers prove nothing, the qualitative result does show that if one reduces demand, output will fall. All variables fell less than employment, some dramatically less. The rise in real income softens but cannot, in any realistic context, reverse the decline. The effects of changes in prices derive from the following percentage changes:

$q_1/p_1 - 19.1\%$
$q_2/p_1 - 10.7\%$
$q_1/p_2 - 3.9\%$
$q_2/p_2 - 6.2\%$.

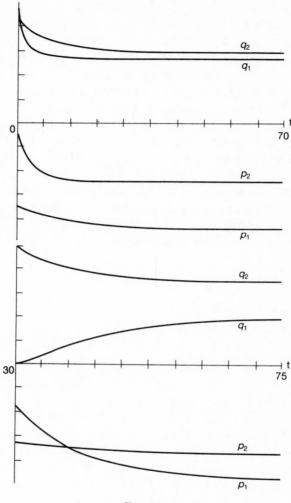

Fig. 1.1

Such large changes in the value-system will, in general, provoke
consequent further changes in production processes.
To simulate this I have assumed that at $t = 30$, sector 1 cuts its
use of good 2 by 50%, i.e. a_{21} drops from 0.10 to 0.05. The
resulting compound evolution is illustrated also in Fig. 1.1.

The results are most interesting: in the previous case all motions were in the same direction, but here there is one dramatic exception, q_1. With both parametric changes operating the percentage changes from $t = 30$ to 70 were as follows:

$q_1 + 1.56\%$
$q_2 - 5.94\%$
$p_1 - 9.68\%$
$p_2 - 0.29\%$.

If one looks carefully at the dynamics of the problem one can see the reason; also one can see how complex is the dynamical behaviour of this extremely over-simplified system. Such an exercise can give one an insight into the enormous complexity of an economy with thousands of sectors and millions of agents, all in varying degrees different. The results even in an almost-linear system can be quite erratic, independently of any random shocks. Thus in spite of a decline of 50% in the parameter at time $t = 30$, but because of the special behaviour of sector 1, employment only fell by 2.66% from $t = 30$ to 70, whereas it fell by 46.3% from $t = 0$ to $t = 30$.

This simple simulation will, I hope, help to convince the reader, if not already convinced, that aggregative models, for example using GNP, price indices, etc., will usually mask the reality, however helpful they may be to illuminate the nature of problems in economics. I say this in spite of the fact that I am now going to discuss chaotic attractors, by means of aggregative models! I do this because the problem is so new, so difficult, so incompletely understood, that multidimensional models are out of the question.

Dynamical systems display themselves as attractors and/or repellors (sinks and sources, or basins and hills). There are motions into or out of *fixed points* (or equilibria). There are also attractors/repellors for fixed rates of change, or also other well-defined motions such as periodics. The interesting and difficult cases are the *chaotics* (more soberly called aperiodics) which are not yet, and possibly never will be, fully understood, precisely because they are ever-changing. Nevertheless there are chaotic attractors, where the motion follows a recognizable pattern, though always changing within the pattern.

In linear systems there are transients which dissipate, leaving the steady state: for non-linear systems, it is not so simple, and initial conditions may produce distinctly different long-runs. Frequent random shocks may create for linear, non-chaotic systems a result that may look much like the behaviour of an undisturbed chaotic system. Given a statistic produced by a chaotic dynamical system which is also subject to exogenous disturbances, if one can correctly guess or otherwise discover the correct dynamical model, then one can extract the proper deterministic-chaotic component from the statistic, leaving the residual as the irregularity due to the shocks. This fact surely makes it urgent for econometricians to consider a wide range of promising chaotic dynamical models.

In a sense it all began about a century ago with Poincaré, who treated what would now be called bifurcation. A bifurcation is a change in the qualitative nature of solutions to a system, and is related to the concept of structural stability. The change may be produced either by the variables of the system itself or by exogenous variation of the system's parameters. The change may consist of a shift from repellor to attractor or from a fixed point to a fixed motion. These are to be sharply distinguished from changes in magnitude or duration of the existing solution.

Consequently it becomes important to determine the domain of attraction or repulsion: this yields a kind of catchment area for any particular type of motion. A single system can have more than one catchment area. Thus a differential limit cycle will ordinarily have an unstable (repellor) fixed point but a global attractor, so that between the two must lie at least one closed boundary which constitutes a stable fixed motion, or limit cycle.

In order to clarify the fundamental paradox of chaotic attractors, I propose first to consider a familiar type of cyclical, differential system. Like meteorology, economics can often yield good short-term prediction (not, however, for the recent crash), not so good for medium-term and more or less worthless for long-term. The mathematician John von Neumann once maintained that, given sufficient funds for global weather data collection and a megacomputer, he could provide detailed, accurate prediction of weather over time and space. He based himself on the fact that, by contrast with economics, we know very well the laws of fluid dynamics. Hence, given the initial conditions globally, we could compute the flow in space and time of the weather. It was good sound methodology and I long believed he was right, though his assertion remained untested because of the scale of costs. I now think that empirically his conclusion was wrong in principle. To understand why he was wrong will, I hope, illuminate how fundamental for applied mathematics was the discovery of chaos and why it has unleashed a flood of new mathematical work. Mathematically, of course, von Neumann made no mistake: if we have a correct model, then, given the initial condition, the sequence of events is necessarily calculable and therefore any future event must be predictable exactly—the entire logical basis of mathematics rests on this.

The rot began with another great mathematician, Poincaré, who, faced with some unpredictable astronomical irregularities, spoke of homoclinic tangles, or non-wandering sets, and doubted that they would ever be understood. The American mathematician George Birkhoff carried Poincaré's work forward by developing ergodic theory. It was one of his pupils, E. N. Lorenz, in the 1960s, who really set the subject in motion—significantly by computer simulation of atmospheric irregularities. His system really posed the basic paradox: three non-linear differential equations with specific initial conditions yielded unpredictable results. This seemingly blunt challenge to mathematical logic has produced an explosion of theorizing, which goes a long way, but not all the way, to resolving the paradox.

Is this merely a storm in a mathematical teacup, or does it raise a serious problem of wide application—in particular to economics? I urge the reader to think about it: how serious it is remains an open question. Scientists have, by taking certain types of issue, produced stunningly successful results. But a large area of natural phenomena is beyond the reach of natural science—the fall of a leaf, or turbulent flow of air, etc. In particular, as we all know, economic time-series are highly erratic, so we must, it seems to me, take these new lines of investigation seriously. It should be remembered that probabilistic analysis is a confession of partial ignorance.

To come more specifically to economics: the Great Depression gave rise to two theories, the Kalecki and the Hansen–Samuelson models. Both were linear but such models are incapable of explaining the continued existence of oscillations. Frisch misled a generation of investigators by resolving the problem with exogenous shocks, whereas already in the 1920s van der Pol had shown (as Frisch should have known) that a particular form of non-linear theory was the appropriate solution. His solution leads to a limit cycle.

For differential systems to give rise to chaotic attractors, the dimension must be three or more. With a cycle model formulated as two first-order equations, a time-varying parameter then raises the dimension to three. Suppose that there is a simple sinusoidal investment function influencing x, giving the system as $\dot{x} = \varphi(x, t)$. x has a natural period of $T_N = 2\pi/\omega_N$. If the forcing function has a period $T_t = T_N$, the result is a torus with a point in state space returning to any position once each cycle. By taking T_t shorter and shorter one gets a succession of points, which can be registered on a Poincaré section.

Proceeding thus one can get a continuous curve for the conjunction of the two cycles. This complicated and confusing experimental result has been much clarified by a Tübingen mathematician, Rössler. One uses a Möbius strip, top to bottom, sides reversed, giving a one-sided surface. For an appropriate T_t, one can analyse the result with a three-dimensional model.

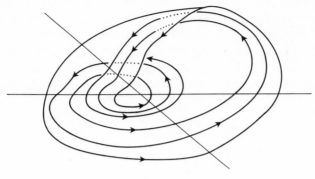

FIG. 1.2

Following a trajectory, an outer loop expands to the right and upward; then folds over and descending contracts to an inner loop and so forth. I give a two-dimensional projection which shows plainly the two cycles (the simulation is from a related model). It will be noticed that the lines tend to get grouped

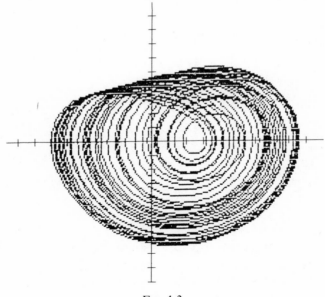

FIG. 1.3

with empty spaces in between; if the simulation is continued
these bands become dense, i.e. ergodic. This result was first
predicted by Birkhoff and first simulated by Shaw.

If the simulation is carried on long enough the dark bands
become solid black. In fact they represent flow lines at different
levels (Fig. 1.4). In my three-dimensional drawing they are

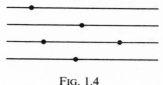

FIG. 1.4

represented as a flat surface but in fact they are very thin
layers, like French pastry. the loops get folded and arrive
on different close layers (Fig. 1.5). They are called Smale
horseshoes. Astonishingly enough these experimental results
have been found to conform to a branch of abstract number

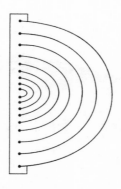

FIG. 1.5

theory—Cantor sets, also now called 'fractals'. Take a line,
divide it into three equal parts, then do the same repeatedly
with the two end-parts. The result is like Smale horseshoes
(Fig. 1.6).

FIG. 1.6

With the foregoing superficial account in mind, one can attempt to resolve the paradox that a deterministic system can be unpredictable. The multiple surfaces expand, fold over, and contract. Consequently one sees that points initially close together and increasingly diverge over time, both in direction and distance travelled. Therefore since one can never determine an exact initial condition in applied work, one increasingly loses accuracy. By contrast mathematically one can state exact initial conditions and calculate precisely the resultant flow, but this is not so for empirical analysis. Secondly, on each circuit the flow goes on a different level so that the resultant dense bands defeat precise detection of subsequent positions.

Chaotic attractors introduce a remarkable, awkward new conception. Thus originally there was the fixed-point attractor, or stable equilibrium, *sensu stricto*. Poincaré generalized the conception of equilibrium to include equilibrium motion in the form of limit cycles. But for chaotic attractors there is neither a fixed point nor a fixed motion. This so violently generalizes the notion of equilibrium that the original conception is more or less lost. Can a system which never repeats itself over time still be considered in 'equilibrium'? It is a bounded motion, a non-wandering set.

We left Schumpeter impaled on a contradiction: his innovation produced decline instead of growth. The aim of his theory was precisely the reverse and that, in fact, is its superiority to all other dynamical and cyclical theories. Our perverse results arose because A, the investment, remained constant. There was some justification for this since his innovational investment had returned to zero on completion, but he, of course, escaped our dilemma by assuming full employment at the new, higher productivity. Our results are in conflict not only with Schumpeter but with the economic history of capitalism. The economy is characterized by growth, not decline, but interrupted by spectacular collapses as in the 1880s, the 1930s, and the 1970s. Therefore particular attention must be given to innovations, or technical progress. This is most interestingly done in a discrete-time, non-linear model: it is a striking fact that such a very simple model can give rise to almost the entire gamut of erratic or chaotic solutions. There are many significant time-lags in economics, for example breeding time for livestock, trade union wage negotiations, or construction time for new capacity, which is the one I shall consider.

To keep matters simple, I take the case of a uniform lag, thus keeping to one dimension.

Consider the case of a group of related, innovatory methods of production or new products. There is considerable agreement amongst economists and economic historians that the introduction of a successful innovation is well represented by the logistic function, either

$$\dot{x} = bx(1 - x/\bar{x})$$

or

$$x_{t+1} = ax_t(1 - x_t/\bar{x}),$$

where $a > 0$ and $\bar{x}$ is the maximum level for the innovation.

Initially it appears in a rather crude form and is unfamiliar; then as it gains acceptance it reaches its maximum rate of expansion. After that there is a slow deceleration of the process of absorption as it approaches the complete integration into the economy. The resulting curve has the shape of an elongated s with a non-negative slope. Making a change of variable to non-dimensional form, with $\bar{x} = 1$, the result is

$$x_{t+1} = ax_t(1 - x_t) = ax_t - ax_t^2.$$

The result is that x maps into a unimodal, or 'one hump' function with its maximum at $1/2$. For $a \leqslant 3$, it is an attractor, and for small a it is a monotonic attractor. Its fixed point where $x_t + 1 = f(x_t)$ is $\hat{x} = 1 - 1/a$. As larger as are considered, the max point rises above the 45-degree line and a stable oscillation arises where $-1 < f'(x) < a$, as is exemplified in Fig. 1.7. For $a = 3$, $f'(x) = -1$ and the result is a neutral cycle in the neighbourhood of $x = 2/3$.

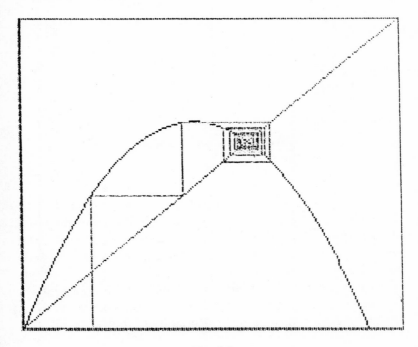

FIG. 1.7

For larger values of a, $f'(x)$ becomes < -1 and the fixed point generates an unstable cycle. This is illustrated in an enlarged section in the Fig. 1.8.

As a is increased towards 4 and $f'(x)$ sinks further below -1, one finds a bewildering collection of all forms of behaviour—oscillators, both attractors and repellors, as well as aperiodic

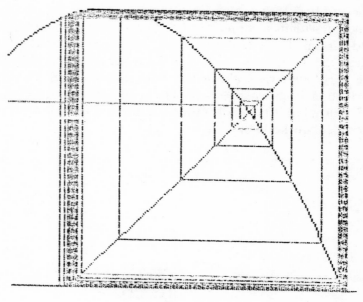

FIG. 1.8

motion, both stable and unstable, all mixed together, in a word, *chaos*. This is illustrated in Fig. 1.9. As *a* approaches 4 the variety of succeeding bifurcations leads to the conclusion that it is helpful to write $f(x, a)$ since the behaviour depends on both the variable and the parameter. This behaviour has been the object of close study and some, but incomplete, light has been shed on it. I quote from Thompson and Stewart commenting on this one-dimensional mapping:

The most ambitious task would be to detect the sets of values of *a* for which the motion is periodic, ergodic or mixing (chaotic) respectively. In this way it would be possible to distinguish.... Unfortunately this is in general impossible, even for very simple maps such as the logistic map, so we must content ourselves with more generic results on the *size* of the subsets. The fundamental result in this field is the following:

Choosing a value of the parameter *a* in the prescribed range there is a non-zero probability of mixing behaviour and a non-zero

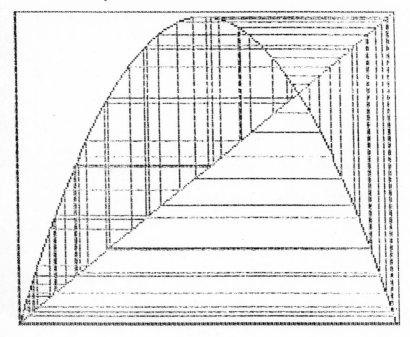

FIG. 1.9

probability of ergodic behaviour. Since it is also clear that periodic motions occur with positive probability we can say that the parameter range, or control space, is partitioned into three subsets whose Lebesgue measures are all positive.

Difference equations lend themselves naturally to being investigated sequentially or *iteratively* and in this way much can be learned of how and why and when they bifurcate and thus mix types of solution. Therefore it is helpful to look at the *iterates* of the function. If $x_{t+1} = f(x_t)$,

$$f(x_0) = x_1$$
$$f(x_1) = x_2 = f(f(x_0)) = f^2(x_0) = f(x_1),$$

hence $f(\ )$ is a different function from $f^2(x_0)$.

Hence the first iterate of $f(x)$ with the appropriate variable must equal or pass through the same point as $f^2(x_0)$.

$$f^3(x_0) = x_3$$
$$\vdots$$
$$f^m(x_o) = x_m.$$

The study of these iterative maps has been intense in recent years and has yielded some insight into irregular or chaotic behaviour. The number of iterates increases very rapidly as a approaches 4; beyond that point arrive the complications of Cantor sets.

FIG. 1.10

For $a > 3$ the number of periods doubles successively as a increases (see Fig. 1.11). The doubling can be seen as a function of the parameter a. The beginning of complications is called period doubling, and then the doubles are each doubled and so on to great complexity.

To illustrate the problem I take an economy in equilibrium, with an innovation in its early but not earliest stage. The innovative capacity, k, grows logistically and its effect on demand is multiplied by m, the output–capacity ratio. The saturation level for any one burst of innovations is n. The current demand for goods, both for input and consumption is d per unit of output, q. Thus demand consists of current demand plus demand from the innovational investment, with

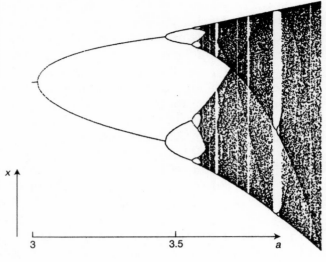

FIG. 1.11

output adjusting by a proportion, f, to any discrepancy between output and demand, thus

$$q_{t+1} = q_t + f(m(ak_t(1 - k_t/n) - k_t) - (1 - d)q_t),$$
$$k_{t+1} = ak_t(1 - k_t/n).$$

Using vaguely plausible parameters, the simulation shown in Fig. 1.12 results. Initial output is taken to be in equilibrium,

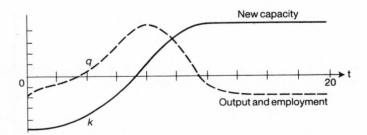

FIG. 1.12

so that as Δk rises, so also does output, with a modest lag. But then as innovation reaches its saturation and Δk drops to towards zero, output falls back to its initial level. Therefore innovation, by itself, stimulates a boom but also a subsequent collapse. Even this conceals its effect on employment (taking the innovation to be labour-saving). If inputs of labour per unit of output have fallen by 20%, so also will employment have fallen.

For the following innovative boom, one may assume a lower employment requirement, but also lower prices, higher real income and consumption. If real consumption per unit of output rises by less than the corresponding fall in labour inputs, there will be a decline in demand per unit of output. Consequently in the succeeding cycle the economy will relapse to a lower level of output and an even lower still level of employment. As the second wave of innovation is exhausted, output returns, not to its previous level but somewhat more than 2.6% lower than the initial value (with employment even lower), though this is not easily seen in the graph of the simulation (Fig. 1.13).

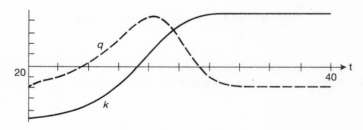

FIG. 1.13

Hence we see that innovation alone is not the explanation for a rising trend to output and employment, if with Keynes, we believe that output is controlled by effective demand. The only basis for the opposite conclusion is that of the naïve view that there is always market clearing for labour, i.e. full employment.

As Schumpeter maintained, at the end of an innovatory burst, a new one will commence (because of the exponential growth of new ideas). Thus a second boom and collapse is initiated and so on for as long as capitalism persists and technological conceptions continue to grow at around 4%. It is important to note that innovation is crucial to cycle theory because, in depression, with generalized excess capacity, there is no other basis for the new investment necessary to generate a new expansion. Hence Schumpeter is rescued by Keynes (though he would never have admitted it).

$\dot{k}$ goes to zero but output does not sink back to its previous level. The explanation for this must be sought in the continuing multiplier effects of exogenous real demands, which may continue after innovatory investment has ceased. In this curious fashion, Schumpeter's pungent dictum is validated: the cycle is simply the form in which growth takes place. Hence, as we all know, capitalism is creative, but in a chaotic way, generating both growth and unemployment. This has never been more evident than now, with the cybernetic revolution in technology in full swing.

for a basic constraint on growth, but that we substitute labour for 'land', i.e. natural resources. By virtue of technical progress, there has been a huge growth of output and of employment without coming up against effective shortages of land or natural resources. What has happened is that various shortages of labour have been resolved repeatedly by labour-saving innovations. This does not imply that it will always be so: we may run short of fuel or we may so reduce the need for labour, by automation, that it ceases to be a constraint.

Thus with a fixed technology and real wage, there is a constant growth rate if Say's Law holds. Thus if $a = 0.2$ and $a_L = 0.7$ with a unit real wage, the growth rate is 11% per year. The profit rate is $(y_{t+1} - y_t)/y_t$ and is equal to the growth rate. In stagnant economies all or most of the profit (or rent) is consumed and the wage is conventional, presumably deter-mined by historical evolution. If this wage is high enough to lead to growth in the labour force, one finds the common situation in the LDCs of unemployed surplus population.

Now consider a somewhat different aspect of an isolated agricultural economy. Assume that the corn production is embedded in a wider economy so that both consumers and producers can switch to alternative goods. Suppose that the demand for corn is a non-linear quadratic:

$q_t = -ap_t + b + cq_t^2 + d$, and that output is linear:
$q_{t+1} = dp_t - e$.

Measuring in deviations from the fixed point (constants zero),

$$q_{t+1} = -(d/a)q_t + (c/a)q_t^2.$$

Such an equation can easily become a chaotic attractor, depending on the values of the parameters and on initial values as well. In view of the many varieties of supply and/or demand curves, it is not unrealistic to investigate various parameter values. With $d/a = 1.05$ and $c/a = 0.60$, a discrete limit cycle results, as is shown in Fig. 2.1, for 40 years and with various initial values.

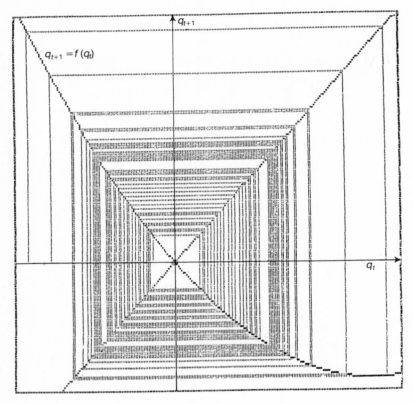

FIG. 2.1

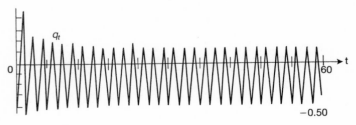

FIG. 2.2

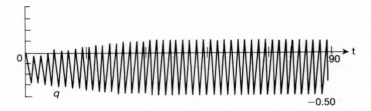

Fig. 2.3

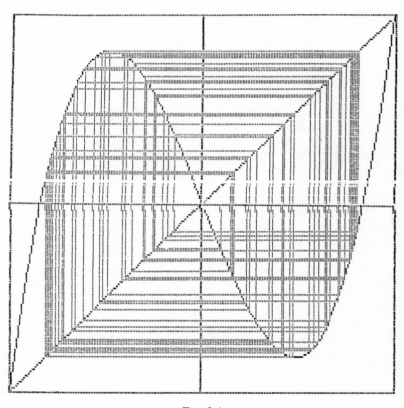

Fig. 2.4

From an inspection of Figs. 2.2 and 2.3, one finds that the limiting values, from inside and from outside, are not identical. Unlike differential limit cycles, these motions approach not a closed curve but an annular band. Once the point has reached the limit band, which it does in finite time, unlike the differential asymptotic approach, it will remain there indefinitely unless disturbed. Different initial conditions result in somewhat different behaviours, so that there is a slight scattering, which destroys precise periodicity. Of course an approximate periodicity is guaranteed by the market structure under the influence of the solar period.

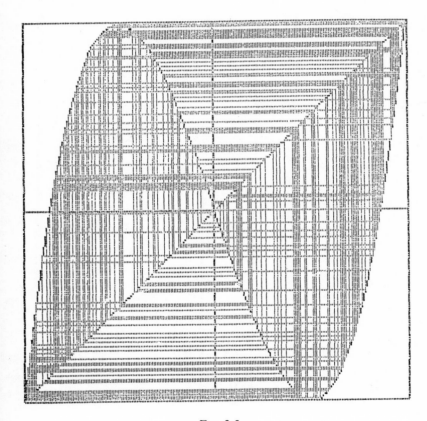

FIG. 2.5

Non-linear supply and demand curves can easily give rise to a quadratic form for $q_{t+1}=f(q_t)$. The dramatic consequences which follow from such a situation were totally missed by all the earlier investigators of lagged markets: the whole concept of deterministic chaos was unknown, in addition to which, in the absence of computers, one could not easily investigate the many possible consequences of all combinations of parameter values.

In Fig. 2.4 is given the behaviour of the first iterate around the fixed point for the equation

$$q_{t+1} = -2.5q_t + 3.5q_t^2.$$

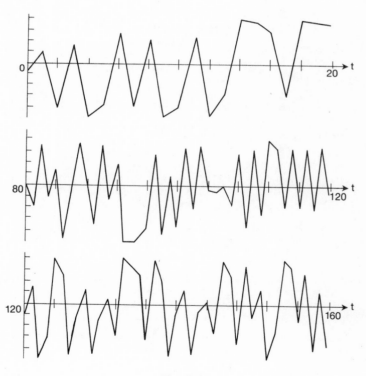

FIG. 2.6

This is highly erratic behaviour indeed and it all occurs in the complete absence of any exogenous shocks.

In Fig. 2.5 is given for the same equation the results with the parameters altered to -3.0 and $+4.0$. Evidently the result is truly chaotic, deterministic behaviour from one of the simplest of all non-linear dynamical models. Of especial interest is the fact that, though it is ever-varying in shape, it never leaves a definite region. One can consider it as a stupendous generalization of a stable motion, as that is a generalization of a stable fixed point. This is why it is called an attractor and it constitutes a non-wandering set of points.

To see just how irregular such a model can be, Fig. 2.6 reproduces the resulting time-series for the years 0 to 20, 80 to 120, and 120 to 160. From statistical analysis there is, as far as I know, no way of determining the structure of the model. More particularly, if this model had been subject to exogenous shocks, there is as yet no way to separate the deterministic from the shock effects. Until now such a series would have been regarded as dominated by shocks.

3

The von Neumann Model
as a Chaotic Attractor

THERE is a remarkable direct line of descent from Ricardo to Marx to John von Neumann. Marx acknowledged his debt to Ricardo, but von Neumann knew no Marx and had no interest in the Marxian analysis. Why, therefore, is there a connection? In my view it is because von Neumann saw more deeply than others into the nature of the economic system and therefore arrived at an analysis with certain parallels with Marx. Be that as it may, he had some conversations with Kaldor in the summer of 1931 in the course of which he rejected neoclassical economics and outlined his conception of how to formulate economics, stating that it should be analysed like a slave economy, something like ancient China! This is the connection with Marx and Ricardo: a fixed real wage.

In 1932 he gave a seminar on his theory at Princeton, with no evident effect. Then he repeated his proposals at the Karl Menger seminar in Vienna, and from that was published in 1938 'Über ein Ökonomisches Gleichungssytem und eine Verallgemeinerung des Brouwerschen Fixpunktsatzes'. It is arguably the greatest single advance in economic theory this century. Since Nobel did not include mathematics in his prize categories, von Neumann could never have received the prize, but surely he could have qualified for the economics prize on the basis of this single work.

With its stunning generality and power, it establishes the existence of the best techniques of production to achieve maximum outputs of all goods at the lowest possible prices, with outputs growing at the highest possible rate. Thus he transformed general equilibrium from statics to dynamics.

Given the real wage, even with a constant technology, there necessarily results a surplus in production (this is the Marxian connection), so that if there is to be dynamical equilibrium of supply and demand, cost and price, the economy must grow at a determinate constant rate with a constant rate of profit/interest. He dealt with n goods, m processes, and n prices, but I can save the reader and myself, much mental stress by dealing with one good and one price, and one real wage—our corn economy (this is the Ricardo connection). He implicitly applied it to a modern economy, but it really fits only a simple agricultural society. Furthermore, for obscure reasons, he formulated it as a lagged system, but this really fits only the common solar cycle, and hence the Classical dynamic. With seed, rent, and wages constant proportions of output, the growth rate is determined, if, as in von Neumann, all the profit/surplus is invested in growing output. The available land of equal fertility is unlimited as is the supply of labour, as in von Neumann. In effect, one produces employment by producing the real wage. The result is thus the same as in the previous analysis (Chapter 2).

Along with its greatness, the von Neumann model has two crucial, basic defects: the infinite supply of labour and the constancy of technology are in glaring contradiction with the realities of the modern, industrialized economy. It is urgent that these two defects be removed in order to apply the theory to contemporary society. The first defect I shall deal with mainly in this section, reserving technological complications for Chapter 4.

The total labour supply grows at a constant rate, g_N: it is therefore unproduced, a rent good like land, and its price is determined by scarcity, not cost. If the wage is low in relation to productivity, then profit and growth rates are high, leading after a time to rising wage rates and falling profit and growth rates. This is a self-regulating, feedback mechanism. It is the simplest way to explain the extraordinary fact that labour supply, employment, and output have multiplied themselves many times over in a century without serious misalignment.

Labour productivity rises at a constant rate: the real wage may rise more rapidly or less rapidly yielding either a rising or falling unit labour cost, u. This in turn will alter the growth rate of output and employment.

The model, with Say's Law, gives

$$y_{t+1} = y_t/(\bar{a} + u_{t+1}), \text{ where } u_{t+1} = (a_L w)_{t+1},$$

i.e. unit wage cost, proportional to the share of wages, $v_t = L_t/N_t = a_L y_t/N_t$. Let $\Delta w/w = f(v) = b(1/(1-v)-c)$, $b > 0$, $c > = 0$. b alters the sensitivity of the wage to the employment ratio, v.

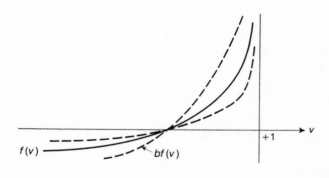

FIG. 3.1

$$u_{t+1} = (1 + f(v) - \bar{g}_a)u_t, \text{ where } \bar{g}_a = -\Delta a_L/a_L.$$
$$\Delta v/v = \Delta L/L - \bar{g}_N = \Delta y/y - (\overline{g_a + g_N}).$$

Hence

$$v_{t+1} = ((1/(\bar{a} + u_{t+1}) - (\overline{g_a + g_N}))v_t,$$
$$u_{t+1} = (1 + \bar{b}(1/(1 - v_t) - \bar{c}) - \bar{g}_a)u_t,$$

along with y_{t+1}.

Assuming a 'period' of one year, Fig. 3.2 gives for reasonable parameters the phase portrait of u and v over 50 years. They

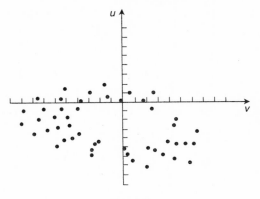

FIG. 3.2

evidently oscillate with a constant periodicity which is moderately stable dynamically. There remains the problem of the inapplicability of a single-lag system to an industrial economy. Perhaps the best justification lies in the fact that it is the simplest way to see the basic logic of a system. It is strange that von Neumann chose to use a lag system in preference to the more familiar differential type.

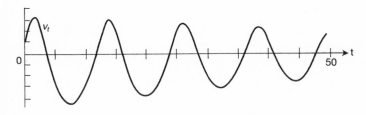

FIG. 3.3

Fig. 3.3 shows what is not evident in the phase portrait— the sense in which a non-linear difference system produces endogenously erratic behaviour. One sees that each cycle is different, though only moderately: from statistical analysis one could not predict turning-points, rate of growth or decline, except very roughly.

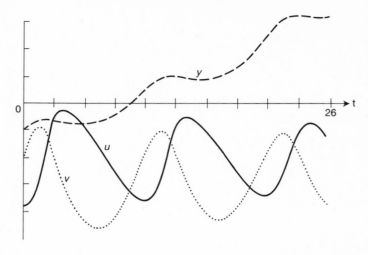

FIG. 3.4

In Fig. 3.4 are simulated just over two complete cycles covering 26 years. It shows that the system generates cyclical growth, with each cycle being moderately different from the other: each peak is different, each trough also, and no two depressions are alike. The long-run average of such a system is steady-state growth at 3%; it grows enough on the average to pay a 2% increase each year in the real wage and to provide employment for 1% more employees. Thus some of the time the economy is growing faster than at 3% and at other times necessarily less, and similarly with wages, part of the time above 2% to be balanced by less than 2% at other times.

The growth rate equals the profit rate, which is first too high and then too low. With a perfect capital market this would also be the behaviour of the rate of interest and the return on investment. Thus all that elaborate, traditional capital theory is unnecessary and gives the wrong answers. This is, of course, a system with no durable capital goods, which von Neumann treats as joint products. Durable goods complicate the problem enormously but also help to obscure certain central aspects of the problem.

Von Neumann's system was dynamically unstable: his fixed-point growth was an unstable equilibrium, so that the slightest deviation from it would lead to ever-further departure. This is serious since unstable motion is never observed, since all actual motion will always depart from it. The foregoing model corrects this fault, only to create another. Since it is stable, in the long run it will cease to oscillate. Therefore it cannot, by itself, explain the existence of cycles.

What is required is a stable, but not asymptotically stable system, i.e. a discrete limit cycle, a closed annular band within which the system will remain forever unless disturbed. What is required is a system which is unstable around its fixed point (equilibrium steady-state growth) and hence will grow in its amplitude of oscillation, but which is stable in its outer region and will experience diminishing amplitude there. A high employment rate, v, means an high growth rate of wage cost or wage share and a low v means a low one. A high growth rate of wages means a deceleration of growth and leads to a decline in the employment rate: the resulting low v means a low growth-rate of wages and restores profit, demand, and output growth, thus bringing the economy slowly back to a high v.

To illustrate clearly the nature of the problem, consider the simplest formulation, i.e. in dimensionless form and in deviations from equilibrium growth. The system in matrix form is

$$\begin{Bmatrix} v_{t+1}/v_t \\ u_{t+1}/u_t \end{Bmatrix} = \begin{bmatrix} a(1.0-v_t) & -1.0 \\ +3.258 & 0 \end{bmatrix} \begin{Bmatrix} v_t \\ u \end{Bmatrix},$$

where a is the unique parameter open to choice and v_{t+1}/v_t and u_{t+1}/u_t each represent unity plus growth rate. The off-diagonal terms being of opposite sign, the system will oscillate. The dynamical stability is determined by the trace (sum of diagonal terms). The one-diagonal term is positive for small v around equilibrium: this produces instability with a growing amplitude. For large v the term becomes negative making the system stable with a diminishing cycle of growth rates.

Consequently between the two regions there must be a barrier since no stable motion can ever coincide with an unstable one. In differential equations this takes the form of one or more curves enclosing the equilibrium point. For difference equations the situation is more complicated and more interesting, as I shall try to illustrate. Enclosing the equilibrium there must exist at least one bounded, annular ring within which the cyclical limit set must lie. The limit set will cluster but not on a single curve because of the irregularity of non-linear difference equation solutions. Since the solutions are discrete points, the approach to the limit set is not asymptotic but occurs in finite time. By an even greater contrast with continuous time, successive points may jump over the ring from outside to inside, but once within the band appear never to emerge.

By considering a series of values for the parameter a, one discovers first a stable cycle leading to a unique equilibrium, fixed point. Then for a higher value the solution becomes a slightly irregular limit set. For a still larger value the limit set becomes sufficiently irregular that it becomes dubious to call it a single closed band. For a small further increase in a one sees that the limit set is imploding inward towards the fixed point. The next step is a completely chaotic attracting region which fills most of the region around the fixed point. Thus one arrives in successive steps to a whole attractive region in which the phase trajectories appear to move in a seemingly random fashion.

Commencing with $a = 2.0$, one has in Fig. 3.5 a very stable cycle leading to a fixed point. Increasing a to 2.8 yields a bifurcation to a limit cycle set (Fig. 3.6). That such a solution generates a moderate but definite irregularity is shown in Fig. 3.7. With a increased to 3.4, Fig. 3.8 begins to exhibit such a marked irregularity that it is difficult to consider it a single closed band. For $a = 3.5$, it is clear that one can no longer consider the set as contained in any closed band; it is expanding in various ways towards the interior (Fig. 3.9). When *a* is increased to 3.6, there is a fully fledged chaotic attractor: the annular band has simply grown to fill much of the enclosed space (Fig. 3.10). The value $a = 4.0$ gives an even more complete filling of the entire region around equilibrium (Fig. 3.11). This means, of course, that the corresponding time-series is highly erratic or chaotic (Fig. 3.12).

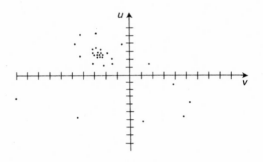

FIG. 3.5

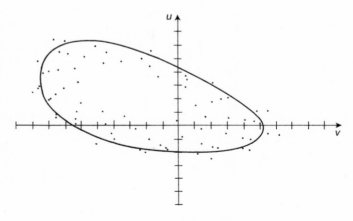

FIG. 3.6

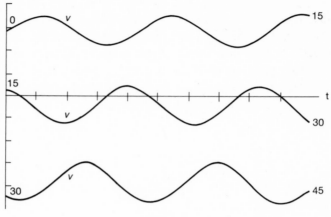

Fig. 3.7

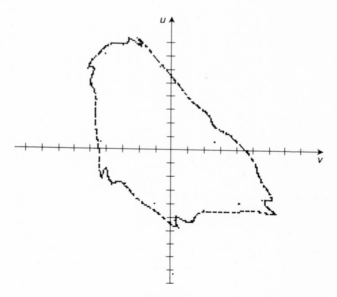

Fig. 3.8

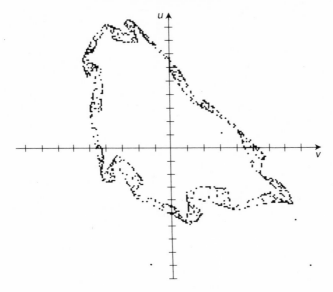

F<small>IG</small>. 3.9

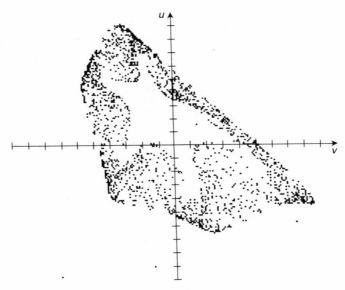

F<small>IG</small>. 3.10

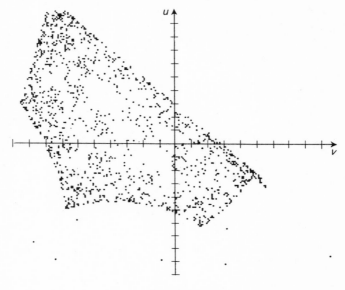

FIG. 3.11

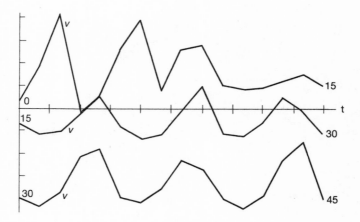

FIG. 3.12

4

Growing in Short and Long Waves: Schumpeter

JOHN VON NEUMANN was a first-class mathematician with little or no interest in social or economic problems. In an inexplicable leap of the imagination he perceived that his highly original game theory could be used to illuminate economics. Significantly, his theory bore at some points a resemblance to Marxian theory, presumably because both probed unconventionally and deeply into the dynamics of an economy.

This brings me to Marx, in my view a profoundly original and important social and economic thinker. In this chapter it is his view that I am following: specifically that the evolutionary nature of economic society is brought about by the perpetual search for profit. He was not trained as an economist, but becoming convinced of the primacy of economic attitudes, he studied the subject and, wisely, became a Ricardian. However, in a fundamental way, he departed from Ricardo by treating profit as a simple surplus which was a crucial and permanent feature of capitalism. Thus he took the best master and ignored completely the newly developing neoclassical marginalism. To give an adequate account of his theory would mean a detour from my present purposes.

Instead, I shall discuss his most eccentric but important follower, Schumpeter. He was much influenced by Marx; whether he arrived at his position independently or whether he was influenced by the two famous Marxists, Otto Bauer and Rudolph Hilferding, who both were in the Böhm-Bawerk seminar with him, I do not know and cannot imagine why I never asked him.

He fully believed in the neoclassical analysis of the essential functioning of the price–market mechanism, but at the same time he departed from this essentially static analysis on the basis of a Marxist analysis. He held that technical progress demanded a thoroughly different formulation which made many of the neoclassical conclusions faulty. Specifically, he concluded that supply and demand for capital was not the clue to the determination of rate of interest/profit: rather it was innovation as a dynamic disturbance which determined the rate, and in its absence the rate would be zero. Like Marx he held that there was an enduring search for innovations which would restore eroding profit. These innovations altered the equilibrium continually, so that the economy was rarely if ever in equilibrium, always moving towards it, though in a cyclical path.

To formalize this one imagines an economy in a state of permanent evolution through technical change. In the boom, high wages reduce profit and growth; this leads producers to urgent search for cost-reducing, labour-saving innovations. The source is new ideas: if we measure new ideas by the number of learned journals in the world, they have increased rather steadily at over 4% for more than two centuries.

Accepting some such fact, there remains the problem of how does such a steady flow get altered into cycles. A plausible analogy is offered by the so-called Roman fountain. Water flows into a basin; when the basin is filled to a certain point, a siphon is primed and it empties the basin at a greater rate than the inflow of water. If the flow in is constant, a periodic outflow of water results: if the flow varies, so does the periodicity.

New ideas arise steadily; when enough have accumulated, a group of innovations are initiated, slowly at first, then with acceleration as the methods are improved and adapted to diverse uses. This constitutes a logistic development and there is considerable agreement that this characterizes the typical trajectory of an innovation. Such a course of events is helped by demand effects: some investment must precede the introduction of an innovation; investment stimulates demand;

increased demand facilitates the spread of the innovation. Then, as all the uses of the innovations are fulfilled, the process decelerates towards zero. This explains what Schumpeter called the 'swarming' of innovations, though he rejected the demand effects, since, in the neoclassical manner, he assumed full employment throughout.

In his final formulation he proposed a three-cycle scheme: Kitchins, Juglars, Kondratievs. The Kitchins were short, well observed, and well understood; they relate to the stocks cycle and innovations play no role.

The Juglars were of the order of 10 years, much studied, but he related them to innovations. He was the first major economist to recognize the serious significance of Kondratiev's study of the historical statistics. He quite correctly related the evidence for long waves to the fact that the most important innovations required 30 to 100 years to become fully integrated into the economy, for example steam, steel, railways, steamships, electricity, the internal combustion engine. He somewhat recklessly applied his innovation theory to both the short and the long waves, and I shall follow him in this.

Lord Kahn in his path-breaking multiplier article formulated the generation of income in terms of a lag, i.e. a dynamic difference equation. Following him, consider an economy with a propensity to spend a proportion $\bar{a}$ out of receipts of q. Reckon in deviations from whatever is happening in the economy—growth or decay, cycle or random variations. What will be the added effect on output, expenditure, and income of an innovative 'swarm' of roughly 10 years? q_t is current rate of output, k_t is new innovative capacity, and $\kappa \triangle k$ is the investment expenditure necessary to create the new capacity:

$$q_{t+1} = \bar{a}q_t + \bar{\kappa}(\triangle k).$$

k_t is in the form of a logistic, i.e.

$$k_{t+1} = bk_t(1 - k_t/c),$$

so that

$$q_{t+1} = aq_t + \kappa bk_t(1 - k_t/c) - \kappa k_t.$$

Such a roughly nine-year swarm is simulated in Fig. 4.1.

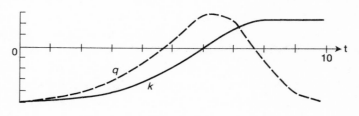

FIG. 4.1

Notice that neither the logistic nor the multiplier is cyclical but that the conjunction produces a wave, and that if each logistic innovative burst is followed by another, there will result a cycle.

Schumpeter assumes that innovational investment produces both the short and the long wave, and implicitly assumes that the two are independent. This poses a complex problem, which he ignored and, at this point, I shall do the same. Assume an existing cycle of about 8 years and ask what would happen if a 50-year logistic occurs in such an economy. Each has a given structure but each is affected by the other.

Given a short cycle of eight years with constant amplitude:

$$q_{t+2} - 1.414q_{t+1} + 1.0q_t = \kappa \triangle k, \text{ with}$$
$$k_{t+1} = 1.15k_t(1 - 0.0020k_t) + 0.080q_t,$$

where the parameter 0.080 allows for the influence of q on k. Taking account of the influence of k on q, one gets

$$q_{t+2} - 1.141q_{t+1} + 0.96q_t - 0.75k(1 - 0.0143k_t) = 0.$$

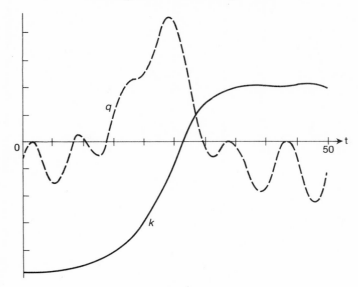

FIG. 4.2

The resulting simulation is shown in Fig. 4.2. The short cycle is completely distorted by the logistic: first it is shortened and made milder; then it is destroyed altogether, becoming merely a slower growth rate. Finally comes the great boom at the point of most rapid innovational investment. After that comes the reverse process but always somewhat different from before. This kind of behaviour could have produced the statistical evidence noticed by Kondratiev, but largely ignored by most economists, except Schumpeter who made it a central facet of his final version of capitalist dynamics. The big ones like steam or steel, would be a part of the short wave but last through several such, with great effect on the performance of each short wave. The mutual conditioning of investment and effective demand helps to explain how a disparate group of technical innovations become ordered into a common upsurge, a case of 'self-organization'.

In principle each wave returns to its initial level and shape

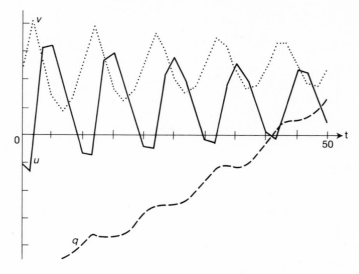

FIG. 4.3

as the logistic approaches saturation level. This, of course, is a serious defect. Schumpeter, like most neoclassical economists, simply assumed labour-market clearing. Hence with improved productivity there must be greater output and income. Once one has understood that full employment is the exception rather than the rule, *lower* inputs mean *lower* demand and output.

To discuss this problem concisely, it is useful to rephrase it in continuous time, with differential equations. In the von Neumann equation of Chapter 3 let q_{t+1} represent a lag, not a time unit. Then $q_t = (a+u)(1 + \dot{q}/q)q_t$, so that $\dot{q}/q = 1/(a+u) - 1$, as before, but in continuous time,

$$\dot{u}/u = \dot{w}/w - g_a = \mu(1/(1-v) - c) - g_a,$$

and

$$\dot{v}/v = \dot{q}/q - (g_a + g_N).$$

Distributive shares between wages and profits are

$$\frac{u}{1-a}+\frac{a+u}{1-a}\,\dot{q}/q=1.$$

With $g_a = 0.015$, $g_N = 0.005$ and plausible values for the other parameters, a typical result is given in Fig. 4.3. It is evident that this gives a growth cycle and thus solves Schumpeter's problem.

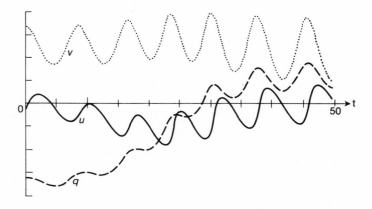

FIG. 4.4

The model is still too simple because it assumes a constant rate of increase of productivity over a 50-year period, whereas the whole point is that productivity only grows in consequence of new technology. Therefore one should rework the model making g_a rise only with the rise of innovative capacity, k. Therefore I take g_a as proportional to $\triangle k$, reaching a peak after 25 years and then gradually subsiding again to zero (g_N is set at zero to isolate the effects of technical progress). This is illustrated in Fig. 4.4. Labour's share falls as investment moves towards its maximum and then returns to its original position: the real wage, however, rises throughout. As average u falls, average v rises, only to fall back again as u rises.

It is illuminating to plot v against q over a 50-year wave. Initially they move up and down in sympathy, but v cannot grow, so that v rises less and falls more than q. Then they gradually return to their original motion. This is represented in Fig. 4.5.

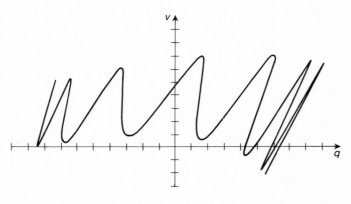

FIG. 4.5

To complete this account I propose to give a simple, abstract view of the basic, essential structure of chaos, in the form proposed by Professor Rössler of Tübingen. It is applicable to a wide variety of problems and is in dimensionless form, but I shall use the three variables, v, u, and k.

$$\dot{v} = -u - k,$$
$$\dot{u} = +v + \bar{a}u$$
$$\dot{k} = \bar{b} + k(v - \bar{c}).$$

The coefficients of v and u being $+1$ and -1, the solutions are necessarily oscillatory and are in deviations from the fixed-point equilibrium, which represents, given constant growth rates of productivity and labour force, a steady-state growth. Setting $k(0)$ equal to zero, or a negligible quantity, $\dot{v} = -u$ and $\dot{u} = v + au$, so that $\ddot{v} - a\dot{v} + v = 0$. If $a < 0$, the result is stable cycle around steady growth. However, with all parameters positive, an unstable cycle results. As *v* grows and becomes greater than *c*, *k* becomes positive; $k(t)$ ceases to be negligible and the full system is operative.

The aim is to illustrate the gradual onset of chaotic solutions as one parameter is varied. Therefore one can set $a = b = 0.2$ and study the solutions as function of parameter *c* alone. With $c = 2.1$ the result is a limit cycle (Fig. 4.6); it is also

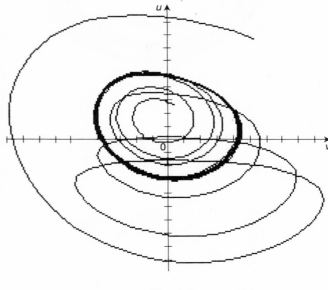

FIG. 4.6

asymptotically stable but in an unusual way. From Fig. 4.6 one sees that the trajectory jumps from outside to inside the limit set, and that this limit set is an annular band, similar to the discrete-time case. Figures 4.6 and 4.7 together show that there is an asymptotic approach by means of successive, slightly different, cycles.

It is possible to give a rough, plausible explanation of the economic meaning of this so-called Rössler Band. It is clear that *u* and *v* can produce a cycle with somewhat realistic

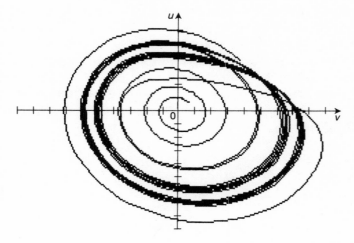

Fig. 4.7

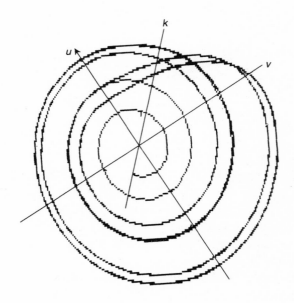

Fig. 4.8

parameters. Then it is equally possible to choose values which give a cycle of growing amplitude. If as a result of such expansive behaviour, investment is heavily encouraged; capacity will accumulate rapidly, more rapidly than output grows, so that investment will eventually decelerate, precipitating a decline in demand and the end of the boom. This introduces a new element with the consequence that the expansion will not repeat the same pattern, leading to a different period and amplitude. Two periods and two amplitudes can coexist as is demonstrated in Fig. 4.8 with $c = 3.1$: the aspect of band width is made even clearer. This is known as period doubling and indicates the onset of chaos. The band consists of a group of similar but not identical cycles. However, it represents not the irregularities of a single cycle but rather similar cycles which keep changing so as to approach a final,

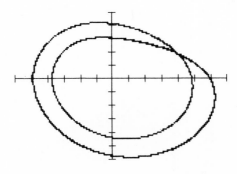

FIG. 4.9

unique limit cycle as illustrated in Fig. 4.9 for $140 \leqslant t \leqslant 170$ and $c = 3.1$. These evident band widths are the initial hint of the possibility of chaotic or aperiodic behaviour. For higher values of c, the number of types of cycles increases rapidly (Fig. 4.10). The trajectories tend to occupy the interior more fully; they implode inward in a manner reminiscent of the discrete-time behaviour. To generate chaos requires three dimensions in differential equations, but the two-dimensional

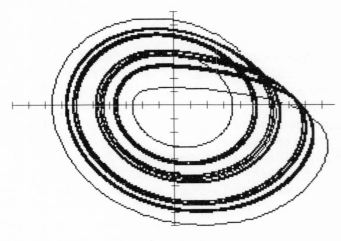

FIG. 4.10

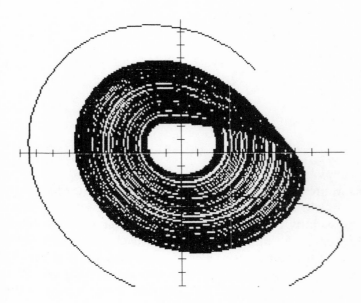

FIG. 4.11

projections exhibit a distinct kinship with discrete-time solutions. The cycles constituting a single band vary only moderately from one another. However, the multiplicity of bands indicates quite unperiodic yet bounded behaviour which is highly irregular. If c be increased to 5.7 (Fig. 4.11), true chaos is present with the ensemble of *almost periodic* bands filling most of the limit-cycle region. For $c = 7.0$ only a small part of the region is left uncovered by a great variety of stable cycles (Fig. 4.12). This obviously produces a chaotic temporal

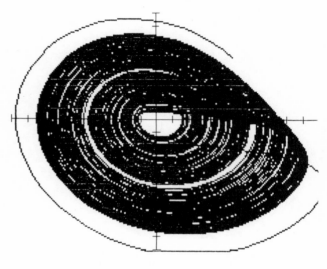

FIG. 4.12

statistic, some indication of which is given in Fig. 4.13 for four successive 15-year periods.

Without independent knowledge one could not extract the model from the time-series, nor could one predict. Thus though the end-state seems quite bizarre, we have arrived there step by step, which helps to lessen the 'strangeness' of our attractor. By varying a single parameter, one illustrates the astonishing generalization of stable equilibrium: from a fixed point to a

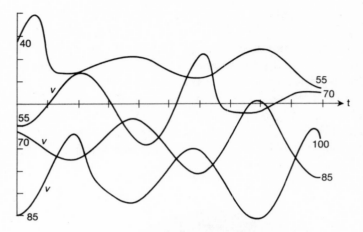

FIG. 4.13

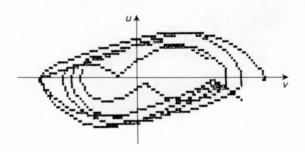

FIG. 4.14

FIG. 4.15

fixed motion and finally to a whole bounded region of non-wandering, non-repeating motions.

One final point: I have not treated a related approach to chaos, i.e. an endogenous cycle subject to an exogenous cycle (see Chapter 9). Thus suppose one has a van der Pol limit cycle generated by an economy, an economy which is subject to a world trade cycle with a different periodicity. This is a non-autonomous model in which erratic, even chaotic, behaviour can arise. In Figs. 4.14 and 4.15 an endogenous limit cycle is subjected to three different exogenous cycles to show the resulting irregular behaviour.

5

The Structural and Dynamical Instability of the Modern Economy

SCHUMPETER, under the influence of Marx, made a major break with neoclassical economics, but under the influence of Walras, accepted most of the rest—in particular the assumption of full employment. Probably this was the dominant reason for his total rejection of the *General Theory* of Keynes.

He formulated economic evolution as technical progress in the shape of 'swarms'. This view, especially stated as a logistic, is finding wide acceptance as an explanatory device in economic history. He said, perceptively, that a technological swarm would shift equilibrium to a new fixed point, but that the path to the new point would not be monotonic but cyclical. But then he blotted his copy-book by asserting that the resulting new equilibrium would be characterized by a higher real wage, higher consumption and output: this is the potential, but by no means the necessary, result, since it ignores the fundamental issue of effective demand as determining output. The innovations have raised the productivity of those employed, but the only firm consequence is lower inputs per unit of output, so that output is left unspecified—more likely than not lower, not higher, than initially, because of the lowered demand for labour and hence consumption demand.

To formulate this, assume output is dynamically adapted to demand; that innovative investment is quadratic over time, producing the accumulation of innovative capacity, k, which reaches 50% of initial capacity after 108 months (9 years). The model is as follows:

$$\dot{q} = e(d + (a + (1 - sk)hw)q + mbk(1 - k/c) - q)$$

$$\dot{w} = f((1 - sk)hq - n)$$
$$\dot{k} = bk(1 - k/c).$$

a_L is labour input/output and the inverse of productivity; $h = a_L(0)$ and $a_L(t) = h(1 - sk(t))$. The parameter s measures the effect of a logistic growth in new innovatory capacity on productivity; b is the growth rate of the logistic; m is the capacity/output ratio; c is the final new capacity level reached in an innovational swarm. Employment, L is equal to $a_L q$ with n as the equilibrium level of employment, at which the real wage is neither rising nor falling. Inter-industry demand is aq and d is a constant exogenous demand. The parameters e and f measure the dynamical adaptations of real output q and real wage w. Parameter values are: $a = 0.2$, $b = 0.12$, $c = 4.0$, $d = 4.0$, $e = 0.161$, $f =$ various, $h = 0.3$, $s = 0.05$, $m = 3.0$, and $n = 2.40$.

There are three essential dynamic elements: the quadratic investment outlay, $mb(k - k^2/c)$; the dynamic Kahn–Keynes multiplication of output; the rate of increase in real wages (and hence in consumption demand) along with the rate of decrease in the demand for labour per unit of output. Even in a crudely simplified aggregative model this poses a formidably difficult problem. In the interest of simplicity I ignore the problematic aspect of accelerational investment as well as explicit consideration of prices. All wages and no profits are consumed and hence enter into demand along with a constant proportion of output: productivity increases proportionately to the accumulation of innovative capacity. A constant exogenous demand of 4.0 is assumed, in order to isolate the analysis of the impact of structural dynamics. The rate of change of the real wage is a proportion, f, of the difference between current employment and that level, say 90%, of full employment, which yields zero change in real wages. Initially wages and output are in equilibrium, and new innovative capacity substantially zero.

In its essential nature, this system is dramatically unstable in both the structural as well as the dynamical sense. As the given innovative investment rises, output also rises; as output rises employment also does, which raises wages and hence demand, thus accelerating the rise in output and ensuring that the demand for labour initially grows fast enough to outweigh the falling demand for labour from rising productivity. There are thus two opposing forces on the labour market, one expansive and one contracting, with their relative strength shifting as the system evolves. The net result at each point of time depends with great sensitivity on the behaviour of the labour market as controlled by the parameter f. To maintain maximum simplicity, the labour supply is assumed constant.

To take the simplest case, assume the real wage constant ($f=0$): the result is illustrated in Fig. 5.1 with $w=1.0$ and

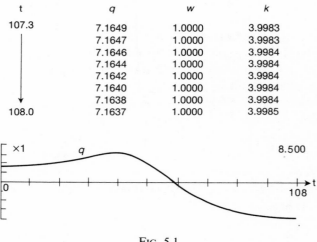

t	q	w	k
107.3	7.1649	1.0000	3.9983
	7.1647	1.0000	3.9983
	7.1646	1.0000	3.9984
	7.1644	1.0000	3.9984
	7.1642	1.0000	3.9984
	7.1640	1.0000	3.9984
	7.1638	1.0000	3.9984
108.0	7.1637	1.0000	3.9985

FIG. 5.1

initial output $=8.0$. Investment rises faster than productivity with the result that output rises to around 8.25, but then the saving on labour begins to outweigh the investment demand so that output begins to fall and continues to fall to its initial value and below, ultimately arriving at 7.16, a decline of some 10.5%. This is, and is meant to be, an extreme case, but it does raise the basic problem of how in detail does demand and

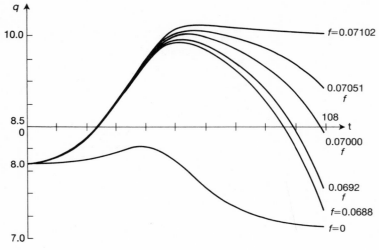

Fig. 5.2

output get determined in the course of substantial technological change.

The behaviour of output as a function of f is given in Fig. 5.2. With $f = 0$, one has the case of the serious decline of output with a constant real wage; raising f to 0.0688 or 0.0692 moderates the decline but still leaves output below its initial value—and falling very rapidly at $t = 108$. An only slightly more buoyant wage, $f = 0.0700$, brings the end point nearly 0.500 above the initial value. With $f = 0.07051$, output has risen nearly to 9.000 and with an optimal wage policy, output rises to a maximum of nearly 10.000 and substantially maintains it ($f = 0.07102$). In Fig. 5.3 the comparative behaviour of wages and output is shown: they move up together but output peaks out at about $t = 60$ whereas the real wage continues to rise though at a decelerating pace. This has, as can be seen, the special effect at $t = 108$, of forcing output to recommence an upward growth; this happens because productivity has ceased to grow whilst leaving employment still somewhat above the equilibrium rate. This special result can be made revealing if we regard the peak output as being full

t	q	w	k
107.3	9.8290	1.6470	3.9983
	9.8294	1.6472	3.9983
	9.8298	1.6473	3.9984
	9.8301	1.6475	3.9984
	9.8305	1.6476	3.9984
	9.8309	1.6477	3.9984
	9.8313	1.6479	3.9984
108.0	9.8317	1.6480	3.9985

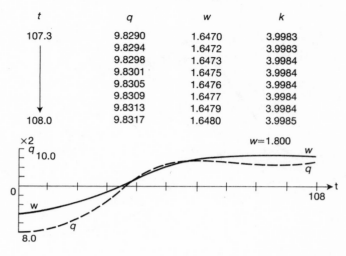

FIG. 5.3

employment; the registered rise in output, being impossible, thus represents nominal, not real, value of both wage and output, i.e. inflation.

It is interesting to consider the level of employment at the end-point, under different regimes. Initial employment was 2.40 and the values reached for various values of f at $t = 108$ are shown in Table 5.1.

TABLE 5.1

f	L	$L/2.40$ (%)
0	1.718	71.6
0.0688	1.788	74.5
0.0692	1.848	77.0
0.0700	2.016	84.0
0.07051	2.14	89.2
0.07102	2.35	97.9
0.07107	2.3	95.8

With output going up and employment per unit of output going down not in step, the balance is a precarious one! The combined effect of the two on demand and output is given by wa_L, which is the unit cost of labour and the share of output going to consumption. That they do not balance in an optimally progressive economy can be seen from the successive values: initially it was 0.30, then at $t = 25$ it was 0.3457, at $t = 50$ it was 0.3869, and at $t = 75$ it was 0.3898.

Governments commonly propose a wage policy based on limiting wage increases to productivity growth, thus implying that wa_L should be constant. This then means a constant multiplier (in this case 2), with the result that output would rise to a peak and then, as the innovations are completed, return to the initial level. The result is that the real wage rises as productivity grows but employment, after a brief rise, declines. The consequences for successive times are shown in Table 5.2. By contrast an optimal wage policy, which ended

TABLE 5.2

t	output	wage	productivity $(1/a_L)$	employment
0	8.0000	1.0000	3.3333	2.4000
25	8.1248	1.0027	3.3422	2.4309
50	8.5634	1.1377	3.7922	2.2582
75	8.2249	1.2381	4.1271	1.9929
108	8.0216	1.2500	4.1667	1.9252

TABLE 5.3

t	output	wage	employment†
25	8.4588	1.1554	2.5309
50	9.7008	1.4670	2.5581
75	9.8232	1.6088	2.3802
108	9.8317	1.6480	2.3596

† Productivity as Table 5.2.

with approximately the initial level of employment, required an expansive wage policy as shown in Table 5.3.

Seen in the light of this exercise, aggressive trade-union wage policy, far from necessarily being harmful, may actually be beneficial—if not too aggressive. Productivity is assumed to be independently proportional to the given logistic growth of innovative capacity.

Schumpeter's conception of economic evolution was that on the completion of one innovative swarm, there would arise another. In accordance with this view, we may take the final values of the first swarm as the initial values of a second one and so forth. Accordingly, initial values are $q = 9.68$, $w = 1.62$, with, as before, $k = 0.025$. Changed parameter values are: $c = 4.7$, $d = 4.65$, $e = 0.07$, $f = 0.09$, $h = 0.24$; these values appeared to be necessary to give a solution qualitatively similar to the first burst. Time runs from 108 to 216, with the results shown in Fig. 5.4. The economy has reached again approximate equilibrium at a higher level of output, a higher real wage with employment restored to its equilibrium value of $n = 2.40$,

t	q	w	k
215.3	12.727	2.3611	4.6977
	12.727	2.3610	4.6977
	12.727	2.3610	4.6978
	12.727	2.3610	4.6978
	12.727	2.3610	4.6978
	12.727	2.3609	4.6978
	12.727	2.3609	4.6979
216.0	12.727	2.3609	4.6979

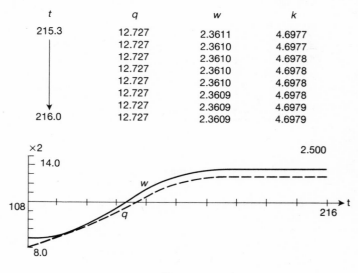

FIG. 5.4

which may be thought to be some constant proportion, say 90% or 95% of full employment. Some of the changes in parameters are obviously necessary, but others seem rather arbitrary; they were chosen as the ones necessary to achieve a new equilibrium at a higher level after nine years. Further innovational swarms could be larger or longer.

The extraordinary instability of the system is both an attractive and a disturbing feature. The slightest changes in parameters give violent results. This is good in the sense that it is a model which can explain unemployment, full employment, or inflation. None the less one cannot credit the real world with quite such a degree of instability and hence one can condemn the formulation as not being robust enough. My intuition is that this is merely an invitation to reformulate along the lines of one or other of the available chaotic attractors.

As I have explained earlier, the nature of such an analysis is that one formulates an unstable (dynamically) system which, however, as it expands, contains a term which grows dynamically and eventually stabilizes the system. This means one gets not a limit point or a limit motion, but rather a limit region within which the system is free to bifurcate back and forth from small, medium, to large motions, thus introducing the crucial erratic feature characteristic of all economic time-series. This vital new point of view is bound to be of great importance to economics and in particular to this problem of structural and dynamical instability.

The aim then is to formulate in one unitary system the whole problem of a long wave with short waves, as a chaotic growth-oscillator. This would give the kind of conceptualization not available to Schumpeter, but one which would adequately represent his original intuition. The simplest and most comprehensible formulation is that of Rössler, which I shall try to cast in economic form. To isolate the analysis of innovational investment and the ensuing structural change in the economy, it seems desirable to deal in deviations from all other aspects of the economy: therefore everything is measured in deviations from the equilibria determined by the vagaries of world trade, fiscal and monetary policy, etc. Wage rates are taken to depend on the ratio, v, of employment to a constant labour force, with $v = 0$ as that level which leaves u, unit labour cost, constant. u is taken to vary by small quantities above and below zero, zero being the level as determined by forces other than innovational.

Since $u = a_L w$, it measures the ratio of the real wage to productivity. If u increases, wages are rising faster than productivity; if it decreases, productivity is rising faster than the real wage. Hence, if innovational investment leads to output growth greater than the growth of productivity, then the real wage will grow faster than productivity during the period of high innovational investment, i.e. $\dot{u} = hv$, $h > 0$.

Also it is presumably realistic to assume that the economy is unstable around equilibrium: for this model, that implies $\dot{u}$ as proportional to u, so that $\dot{u} = +hv + fu$, $f > 0$. Thus u being the share of wages (and salaries) and $1 - u$ is the share of profits, of saving, and of investment (in this simple model). Consequently, the rate of change of output and employment, v, is negatively proportional to u, i.e. $v = d(1 - u) = d1 - du$ (the constant being dropped).

These two differential equations are sufficient to create an unstable economic cycle, as one sees, writing

$$\ddot{v} = -d(hv + fu), \text{ giving } \ddot{v} - f\dot{v} + dhv = 0,$$

as is shown in Fig. 5.5. The instability results from the contradictory action of wage variation: on the one hand a rise in wages increases demand and hence output; on the other hand it reduces the share of profits thus inhibiting the growth of output.

In this manner one creates the type of problem which the Rössler Band is designed to solve chaotically. His formulation of the theory of chaos makes it clear that this new conceptualization is, in a sense, a monumental elaboration of Poincaré's original discovery of the limit cycle. The limit cycle itself is fundamental to dynamics since without it there is no

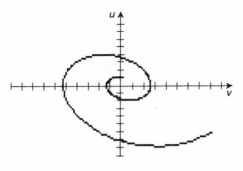

FIG. 5.5

explanation of the continued existence of cycles, as well as the origin of many of them. The Poincaré concept was that, given a system unstable around equilibrium, there must arise a level of one or more of the variables, at which a non-linearity occurs, converting the local instability into global stability. Consequently there is necessarily one or more closed curves (the limit cycles) which separate the region of stability from that of instability.

The Rössler model makes a simple variation on this principle, but one which has an astonishingly rich variety of consequences. When output or employment expands beyond a given level, it induces, not a non-linearity producing a bifurcation from instability to stability, but rather it sets in motion the growth of a variable control parameter which progressively inhibits the expansion of the system variables of output or employment. Thus it does not generate a system bifurcation at a particular static level, but rather one in the control parameter, which has the consequence of creating a band in state space, with an outer and an inner bound. Within that Rössler Band, the system is free to vibrate within a wide range of motions from a slightly irregular to a wildly chaotic fashion: it can vary amplitudes, upper and lower turning-points, as well as periodicities, all in a seemingly random fashion, though in fact completely deterministic. This kind of chaotic attractor betrays its kinship to the original Poincaré conception by being dynamically stable to the Rössler Band in the sense that, if it is initially outside, it will ultimately end up inside, and if initially inside, it will remain inside.

Therefore one defines a third variable, z, as control parameter to the other two variables, with which it is dynamically coupled, thus

$$\dot{v} = -du - ez$$
$$\dot{u} = +hv + fu$$
$$\dot{z} = +b + gz(v - c).$$

By a careful choice of parameters and initial conditions, this single system produces over a 50-year interval approximately five cycles, each different from the other, along with a central section markedly more expansive with a distinctly higher growth rate. The resulting phase portrait for v and u is shown in Fig. 5.6. The time-series for v and u are given in Fig. 5.7, with that of z in Fig. 5.8. It will be evident that the share of wages declines on the average during the higher growth period, only recovering towards the end as relative calm returns to the economy. The variable control parameter, $z(t)$, remains at a low level when the economy is not very unstable, but enters into a period of sharp intervention when the activity level gets high. To demonstrate that this is an attractor, two different

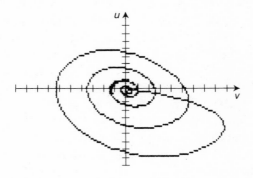

Fig. 5.6

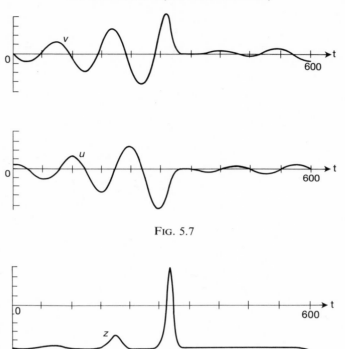

FIG. 5.7

FIG. 5.8

initial conditions are specified in Fig. 5.9, the one being initiated outside the bounded region and the other inside. The one which commences inside remains inside and will never escape, whilst the one beginning outside, after some highly chaotic gyrations, finally arrives in its final resting region, never to depart.

As is evident from Figs 5.6 and 5.7, neither the degree of employment nor the share of labour are periodic, nor do they repeat any single temporal shape, so that one is justified in maintaining that this system exhibits the kind of irregularity characteristic of economic statistics. Indeed, this fact appears to pose a serious problem for econometric methodology: it would be difficult, if not impossible, with traditional

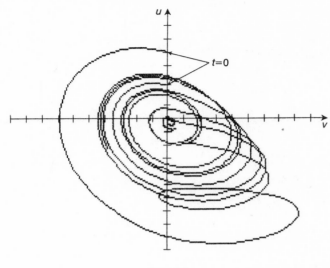

FIG. 5.9

procedures to arrive at a correct estimate of the system's parameters. However, chaos is a generic term covering a wide range of types and degrees of irregularity. In the context of the whole of chaotic attractors, this long- and short-wave system exhibits a rather mildly erratic behaviour. In spite of the irregular appearance of the time-series, the phase portrait for 50 years gives a distinct impression of something like a combination of a short and a long cycle. This superficial view is confirmed by Fig. 5.10, in which a run of 100 years produces approximately ten short cycles along with two long cycles, all with the same parameters and initial conditions as in Fig. 5.6.

By contrast, changed historical conditions, giving different parameters and/or initial conditions, can produce wildly irregular behaviour. Thus merely altering c from 0.004 to 0.005 and the initial condition of v from 0 to 0.01, yields the phase portrait and time-series of Fig. 5.11 for a 100-year sequence; further variation could fill nearly the entire attractive basin with phase trajectories. Hence while the simplicities of Fig. 5.6 might encourage the hope of an early solution to the problem

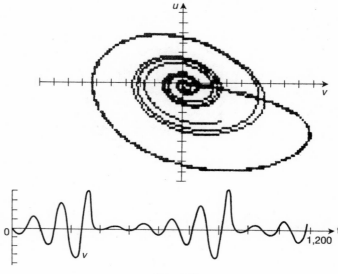

FIG. 5.10

of statistical estimation of such non-linear models, Fig. 5.11 must suggest some doubts. Social scientists are at a grave disadvantage compared to natural scientists, who have the possibility of using controlled experiments to determine their parameters.

The foregoing system shows no growth, being constrained to a bounded region of state space. However, being formulated in terms of ratios, it is compatible with unbounded variables. Though it may not be immediately evident, this system constitutes a growth oscillator as well. To exhibit clearly the fact that this model implies a growing output, consider the following: since v is proportional to employment, L, $\dot{v}/v = \dot{L}/L$, but since $L = a_L q$ (the product of output and productivity), $\dot{L}/L = \dot{q}/q + \dot{a}_L/a_L$, giving $\dot{q}/q = \dot{v}/v - \dot{a}_L/a_L$. With the logistic innovatory swarm yielding a maximum rate of decline in labour input, i.e. maximum growth rate of productivity, in the middle of the Kondratiev, this means that while $\dot{v}/v$ fluctuates above and below its zero equilibrium, $\dot{q}/q$ will rise strongly in

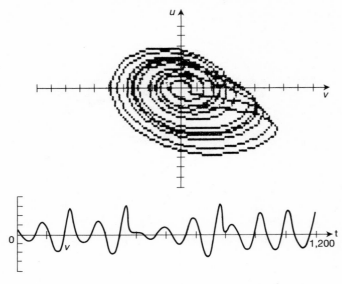

Fig. 5.11

the middle with a positive average value, with the consequence of an irregularly rising output—which will only cease with the termination of the Kondratiev.

To simulate this irregular growth of output it is necessary to add to employment, v, an equilibrium value in order to avoid division by zero. Setting this equilibrium value at 0.9 and assuming the rate of growth of productivity to be proportional to the rate of growth of innovative capacity, the result is shown in Fig. 5.12. In 50 years output grows to somewhat less than double its initial value of 5.000 in a series of strikingly irregular alternations of growth. The onset of a major swarm of innovations leads initially to an almost uninterrupted growth, which, however, suffers two mild setbacks before 25 years. This is succeeded by a vigorous resumption of growth, only to be soon interrupted by a short, sharp decline, followed by a renewal in the form of slow growth, ending with a final mild cycle.

This model of a chaotic growth-oscillator is no doubt neither

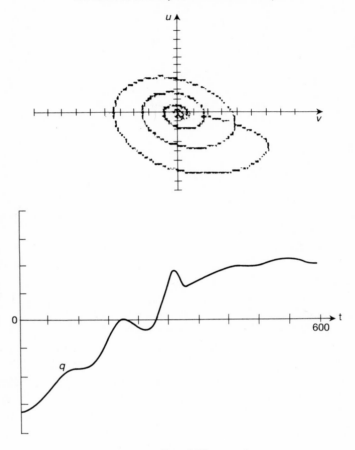

FIG. 5.12

complicated enough, nor realistic enough, nor the best possible cycle model. Its importance lies in the endogeneity of its erraticism, but this in no way denies the importance of the large array of exogenous disturbances which likewise induce erratic behaviour in the economy. What I am suggesting, however, is simply that some such conceptualization of the central generating structure of economic evolution is surely necessary in order to confront the realities of economic behaviour.

6

An Analysis of High and Low Growth Rates†

THE modern industrial economy is so constructed that if there is insufficient growth there will be depression and unemployment. As a result economic analysis tends to be obsessed with growth: it is therefore apposite to consider various aspects of high, medium, and low growth. However, after roughly two centuries of population growth, the prediction is that the more developed countries will in the future experience approximately zero population growth. It is therefore of interest to consider low, stationary, and declining labour forces and the measures to accommodate them.

An adequate treatment of technological transformations in the modern economy requires a multisectoral, multinational system, which becomes immensely complicated. Here I shall consider only an aggregative, single-country version, which will, I hope illuminate some aspects of the problem. A complete treatment would require the resulting changes in relative outputs, relative prices, and hence the consequent shifts in the patterns of world trade.

The fact to be investigated and, if possible, partly explained, is the relatively large spread of growth rates in different countries, in spite of a rapid and relatively free growth in world trade. This occurs in the context of a substantially free, worldwide availability of the elaborate modern technological know-how. If one divides countries into the two groups, MDCs and LDCs, one would not expect similar results as between the

† This section has been stimulated by and benefited from a forthcoming book by J. A. Cantwell.

two groups, but the really surprising fact is the considerable divergences within each group.

It is plausible to seek an explanation of the dispersion of growth rates in the strength of the Schumpeterian conception of innovatory activity. The urge to change forms of production and consumption is obviously economically, but also socially and politically, conditioned. The enormous range of technologies available is either widely applied or largely ignored depending on local attitudes to the restless search for profit.

To model this in simplest form, isolating some essential features, I shall treat only investment in innovational capacity, ignoring all other exogenous real demands, for example the effects of non-innovational investment, government expenditure, and foreign trade. With q as aggregate output and k as innovative capacity:

$$\dot{q} = e(m\dot{k} + (a + a_L w)q + E - q) \qquad 1 > e > 0, m > 0,$$

where m is the capacity–output ratio, w the real wage, a_L input of labour per unit of output, and E all other real demands. Setting $E = 0$, all variables are measured in deviations from their values as determined by E. Thus the model is the same as in Chapter 5 except that $\dot{k}$ is taken as a given, exogenous parameter. Countries with a strong urge to innovate will have a high $\dot{k}$ and those with little such urge will have a low one. The innovative urge may be measured by $m\dot{k}/q$, i.e. the proportion of output devoted to innovative investment. Assuming that the most essential, scarce, unproducible resource is labour, the crucial problem becomes the relation of the growth rate of output to the labour force. It therefore becomes necessary to divide economies into two broad groups: (1) the LDCs with excessive labour supply and (2) the MDCs.

(1) Less developed countries. With their catastrophic demography there is no possibility, for the foreseeable future, of labour shortage. On the contrary they appear condemned to low growth rates with huge and growing populations. The potential technological transformations are enormous, but for

lack of infrastructure as well as social and political conditions, the potential is little exploited. There are exceptions such as Brazil, but even there no threatening labour shortage. Other exceptions are the Asian periphery, S. Korea, Taiwan, Hong Kong, and Singapore (Japan taken as an MDC). They have exhibited strong innovative force, made possible perhaps by concentrating on a limited range of exports, which would allow import substitution for infrastructure.

(2) More developed countries. The wide spread of growth rates here provides the really challenging problem in view of the fact that roughly the same technology is available to all. The basic dynamic elements have been stated in Chapter 5. The rate of innovative investment determines the rate of growth of demand and hence output, as well as, subsequently, the level of each. With a lag this determines productivity and hence a declining level of employment per unit of output. The net effect of these two opposing influences will determine whether employment is rising or falling and at what rate. If the resulting rate is greater than that of the effective labour force, sooner or later labour becomes scarce and its real wage tends to rise, further increasing effective demand and output.

In Chapter 5 the growth rate of the labour force was treated as a constant, exogenous parameter; that is too simple. The population in the MDC countries generally is predicted to be approximately constant with possible future mild decline. But this ignores a decline in work-force because of declining entries deriving from previous falling birth rates and rising rate of exits due to the rising age composition (which is subject to raising or lowering the age of retirement). Then given the number of workers, there remains the number of hours per week including amount of overtime and 'moonlighting'. Of great importance is the variation in the participation rate of women in the labour force; it has shown considerable variation with a strong tendency to increase. This rate is partly a socially given parameter but is also a variable dependent on the female wage rate and the degree of equal treatment of women. Finally there is the role of migration which has played an important part in many MDCs, for example the US, Germany, France, Switzerland, and which is likely to create an even greater problem in the future, given the violent demographic contrast between the two groups. With all these effects, there can be no single prediction of the values of $g_N(t)$ for the group of MDCs or even for any particular country.

The rate of growth of output is given by

$$\dot{q}/q = e(m\dot{k}/q - 1 - (a + a_L w)),$$

i.e. share of investment minus share of saving is proportional to the rate of growth. Since the share of labour is $u = a_L w/(1 - a)$, this may be rewritten as

$$\dot{q}/q = e(m\dot{k}/q - (1 - a)(1 - u)),$$

which says, the share of investment less the share of profits in net output (all profits being saved) determines the rate of growth. A crucial problem is the effect of output on employment, L, and on unemployment, $N - L$, where N is the labour force, i.e. the number of people seeking, or prepared to accept, employment. The employment ratio is $v = L/N < 1$. $\dot{v}/v = \dot{L}/L - g_N$, where $g_N = \dot{N}/N$. Since $L = a_L q$, $\dot{L}/L = \dot{a}_L/a_L + \dot{q}/q$, so that

$$\dot{v}/v = \dot{q}/q - (g_a + g_N),$$

where $g_a = -\dot{a}_L/a_L$, productivity growth. The employment ratio v must remain within a reasonably confined range; it cannot surpass unity and too great a fall can cause social breakdown: therefore it must reach a steady state somewhere within these limits. A realistic, empirical analysis is beyond the scope of this essay, so I shall provide only simple examples to illustrate the nature of the problems. In any case aggregative models are inadequate to give a proper account of the relative shifts involved in structural change and growth.

Innovation appears to have little specific effect on the input of goods, therefore a may be taken as a constant, 0.30. The principal effect is on a_L, which I take to be proportionate to new innovative capacity, so that $\dot{k}/k = g_a = -\dot{a}_L/a_L$. As a_L descends there is a common tendency for the real wage, w, to rise equally, so that as a rough approximation one may take $a_L w$ equal to a constant 0.5. Assuming a constant capacity/output ratio of $m = 2.5$ and an adjustment coefficient $e = 0.8$, one has

$$\dot{q}/q = 0.8(2.5\dot{k}/q - 0.2), \text{ or better, since } \dot{k}/q = m\dot{k}/k,$$

$$\dot{q}/q = 0.8(6.25\dot{k}/k - 0.2).$$

Either $\dot{k}/q$ or $\dot{k}/k$ may be taken as a measure of the strength of the drive to innovate in a country. To achieve a steady state growth rate $\dot{v}$ must be zero (leaving v indeterminate), thus requiring $\dot{q}/q = g_a + g_N = \dot{k}/k + g_N$. Calling 6% a high growth rate, if g_N is +2%, new innovative capacity must grow at 4%, if g_N is 0%, $\dot{k}/k$ must be 6%, and so forth. Mid-range growth being 3% requires a $\dot{k}/k$ of 2%, for $g_N = 1\%$. For a low growth rate of 1%, a labour force growth of 1% permits no innovative investment; only a constant or declining labour force would permit innovative investment. In fact the innovative effort determines both output and productivity; it must also determine in this model the rate of growth of the labour force if it is to maintain a constant average ratio of employment to labour force, for example $g_N = 4\dot{k}/k - 0.16$.

The difficulty is that $\dot{k}/k$ is best regarded as a given fact, only slightly open to control. But the same is true of g_N which is only slowly influenced by public policy or economic performance. Therefore one cannot expect the above dynamic equilibrium condition for constant v to be smoothly achieved. Hence one may seek a flexible adjustment mechanism in response to increasing or decreasing unemployment. I assume it is the real wage rate (via the money wage and prices). When employment is below, say $0.90N$, w rises less than productivity, raising profits and accelerating the growth rate; conversely when $L > 0.90N$, $\dot{w}/w > g_a$, raising wages and lowering profits, decelerating investment and growth.

To analyse the range of growth rates of output, it is necessary to decide what is the principal cause. The source of the differing rates is presumably to be found in the attitudes towards the development of new technology and the amount of research and development. To see the effects of three broad levels of innovative investment, it is necessary to consider the whole economy including non-innovative investment. If all wages and no profits are consumed, the profits all become saving and conventional investment, with innovative investment financed either by the banks, or by activation of idle funds. Setting conventional investment at $(1-u)q$, $1-u$ being the share of profits and treating the share of innovational investment, $m\dot{k}/q$, as a given constant parameter, then

$$\dot{q}/q = e((1-u)+a+m\dot{k}/q-1).$$

Taking a constant labour force as the most general, future condition,

$$\dot{v}/v = \dot{q}/q - g_\alpha, \text{ where } g_\alpha = \dot{k}/k = m^2\dot{k}/q.$$

With $v^* = 0.9$ as that level of employment which equates wage growth to productivity growth,

$$\dot{u}/u = c(v-v^*).$$

The dynamical system can then be written as

$$\dot{v}/v = -bu+f.$$
$$\dot{u}/u = +cv-d.$$

Thus it generates a cyclical behaviour, which suits the alternating character of all capitalist economies: investment is high when growth is and growth is high when investment is. The size of the share of innovational investment, $m\dot{k}/q$, determines whether there will be low (C), middle (B), or high (A) growth of output and of productivity, g_α. The economy does not approach equilibrium, v^*, smoothly, but fluctuates around it: thus v never is constant; rather it is first too high and then too low. None the less, on the average over the whole cycle, equilibrium results: i.e. $\hat{v} = d/c = 0.90 = v^*$. Three different parametric values of $m\dot{k}/q$ produce the three broad types of growth. Setting $a = 0.45$, $e = 0.5$, $c = 1.03$, $d = 0.927$ and letting $f = ea + (e - 1/m)(m\dot{k}/q)$, by choosing three plausible values of $m\dot{k}/q$, one arrives at the following high, middle, and low growth rates:

A with $f = 0.2522$
B with $f = 0.2386$
C with $f = 0.2318$.

The system produces average values of $v = 0.90$ and

A with $u = 0.504$
B with $u = 0.477$
C with $u = 0.464$.

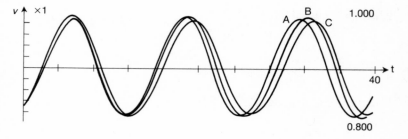

Fig. 6.1

These parameters are vaguely plausible but used only to illustrate qualitatively the types of behaviour. In Fig. 6.1 are given the three time-series for v. They are similar and only slightly erratic, with a moderate lag increasing with time.

In Fig. 6.2 are given the three time-series for the share of wages, where a distinct difference in level is evident. Also in

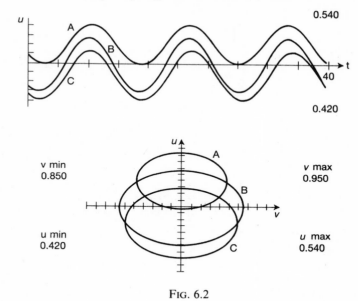

F<small>IG</small>. 6.2

Fig. 6.2 are the three trajectories in state space. The most striking fact in both diagrams is that the share of labour is higher, the higher the growth rate. This may in part be due to the particular choice of parameters, but it does indicate that such a result is possible.

Figure 6.3 shows the dramatic results of the different paces of technological progress; it also exhibits more sharply the degree of irregularity in the time-series. The fact that growth becomes negative is a possible result but not a necessary one: with different parameters there need only be cycles of positive growth rates. Such differences in the continuing growth of output and productivity have considerable significance for the changing structure of world trade and, in particular, for the persistence of balance-of-payments problems.

The behaviour of this model indicates clearly certain central aspects but it is too regular to suggest economic reality. A great variety of exogenous shocks would introduce much irregularity, which can only be introduced for specific cases and

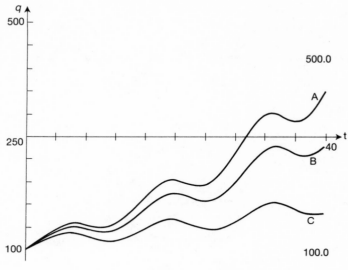

FIG. 6.3

types of problem. I shall only treat systematic, or endogenous, irregularity due to the non-linear nature of the structure. To do this I shall use the Rössler model with three different constant rates of innovational investment along with the resultant growth of output.

$$\dot{v} = -du - iz + fv + e$$
$$\dot{u} = +hv - j$$
$$\dot{z} = +b + z(v - c)$$
$$\dot{q} = +av(\exp(st)).$$

The parameter e represents the level of innovational investment as a proportion of total output; for case A it is 3%, for B it is 2%, and for C it is 1%. Similarly the parameter s is set equal to the rate of growth of innovational capacity: 0.32% for C, 0.64% for B, and 0.96% for A, thus representing three rates of growth of productivity. The other parameters are: d, 0.8, i, 0.8, f, 0.2, h, 0.5, j, 0.5, b, 0.2, c, 4.7, and a, 2.0.

The resulting behaviour for slow, medium, and high growths

are given in Figs 6.4, 6.5, 6.6, and 6.7, including phase portraits. It is evident that the cycles never repeat and also vary according to the rate of innovation. There are approximately $5\frac{1}{2}$ cycles over the 50-year stretch. The endogenous irregularity is modest but definite in both u and v. Each cycle has a different period, as well as often a different shape. Not surprisingly this erratic character varies with the intensity of innovational activity from A to B to C. All three cases have identical initial conditions. Further differentiation would have occurred had the initial conditions been different, since non-linear dynamical-equation solutions are altered by initial conditions.

The foregoing analysis has aimed at illustrating the dramatic effects of various rates of technological progress, the importance of which cannot be overestimated. The post-war period witnessed a rate of increase of output and income which was greater and lasted longer than ever before. Understandably the general public has ardently embraced the vision of endless growth. Likewise economists have placed growth at the centre of their analysis, adding the view that without growth one would have unemployment and depression.

That this growth has brought great benefits to the industrialized nations is beyond question. However, as the process has continued, increasing doubts have been expressed as to the wisdom and necessity of an endless increase of production.

In economics we teach that given tastes and techniques, the economy should be so ordered as to achieve the greatest possible satisfaction of those tastes. But in fact, increasingly one cannot take those tastes as 'given', other than by producers for their own profit: they are being vigorously generated by an increasingly aggressive assault on the mentality of the ordinary consumer by the sophisticated and powerful media, i.e. television, newspapers, etc.

In view of all this, the future of the modern economy should be a matter of great concern and serious, open discussion. To discuss the available options one must consider man-hours of labour rather than simple numbers of employees. The number of workers is not easily altered whereas, in principle at least,

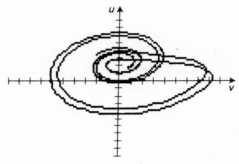

FIG. 6.4

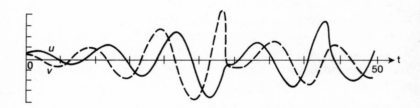

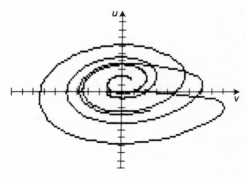

FIG. 6.5

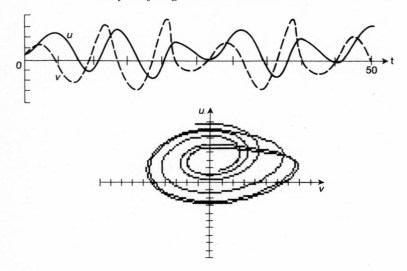

FIG. 6.6

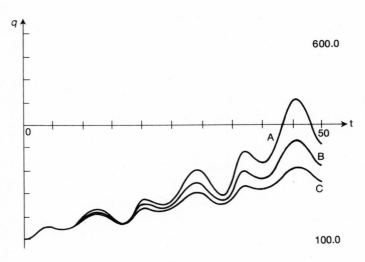

FIG. 6.7

the number of hours worked is highly variable. Therefore I propose to define g_N as the number of man-hours and take it as a variable open potentially to choice.

The essential elements of the problem are as follows: an increase in innovational investment increases $\dot{q}/q$, which in turn increases $\dot{v}/v$, but also by increasing g_a reduces $\dot{v}/v$; any change in v alters the share of wages and consumption with effect on the growth rate of q. Any long-term policy requires $\dot{v} = 0$ with v at some desired level. This then leaves the simpler relation $\dot{q}/q = g_a + g_N$. For any particular society there are persuasive reasons for maximizing g_a, which may require considerable public funding for research and development and technical education. Consider the results of government wartime expenditure on nuclear research, cybernetic control, and robotization.

The relation between growth in output and labour supply is a large and complex issue. The eternal getting and spending is parallel to producing and selling—now heavily influenced and distorted by the search for profit through advertising and the media. There surely should be wide-ranging public and governmental discussion of future policies. For example, by public funding the corruptible image-making industry could possibly be diverted to persuading the public of the attractions of earning less, spending less by working less, without in any way inhibiting rising time-rates of pay through technical progress. Both for government and the public, there exists a range of choices between low, zero, or negative growth rates of output and labour use. This would allow a full and unprejudiced consideration of such things as pollution, exhaustible resources, ambience, quality of life. Thus if $\dot{q}$ is set at zero, then $g_a + g_N = 0$; working hours must fall at the same rate as productivity grows, instead of wages rising at that rate. The real wage per hour rises while the real wage per person remains constant. It might even be possible to arrange, at some cost, for individually variable choice as to the length of time at work. The important thing is that there should be free and uninfluenced choice by the public as to what to buy and how long to work.

7

Irregular Waves of Growth from Structural Innovation

MY aim is to fuse the insights of this century's two great innovators—Schumpeter and Keynes. Each admired the other but failed to incorporate both of their separate contributions. To combine the essential elements of these two original thinkers is not easy and must be rather schematic. I require of the model that it be locally unstable, globally stable: that it generate continuing morphogenesis in the form of structural change in short and long waves which are aperiodic and growth, not stationary, waves.

When Schumpeter said that the cycle is simply the form growth takes, his intuition was a generation ahead of his contemporaries, i.e. the equilibrium theorists, the cycle analysts, the growth-men. His perception of reality was also, alas, totally unrelated to his sadly deficient mathematical capability. I tried to teach him how to use linear cycle theory but he never succeeded in being able to deploy it. When he came to make his final statement on economic development, he opted for three cycles: three Kitchins to a Juglar, three Juglars to a Kondratiev. The very short cycle is important, well understood but involves no innovation, so I shall omit it as a detail. The Juglars and Kondratievs cannot be treated in such a fashion; each is bound to be influenced by the other. Even if they were linear cycles, they would be coupled oscillations and hence both would contain both cycles. With a more complicated, non-linear dynamic, it is necessary to fuse them into a unified model.

Schumpeter, an admirer of the earlier work of Keynes, astonishingly, totally rejected the *General Theory*; for this he may have had two reasons, one bad and one good; first, he believed in market-clearing, including the labour market; the second reason was the formulation of Keynes in terms of aggregate demand, output, and income. The first seems a grave error, but the second poses a basic difficulty. Schumpeter rightly insisted that innovative technical progress is specific to particular industries: hence aggregation masks, and can even falsify, the consequences of structural change. Therefore the analysis should be in large, multidimensional systems, but, alas, I am forced to present an aggregative model, since I am not able to deal with large non-linear systems of the kind I shall be using. My only defence is that the problem and the model are complicated and that a simple version does effectively illuminate much of the dynamical structure. Also it is relevant that precisely aggregate demand is the potent agent of self-organization. Innovations are many and different as to timing and duration of integration into the economy. Consequently, for Schumpeter's theory, the innovative 'swarms' would be so many, so disparate in timing, amplitude, and duration that his cycle would tend to be nearly invisible. But because the level and growth rate of demand plays so great a role in productive decisions—especially in the case of new and risky projects—various innovations are launched and/or rapidly expanded in a rising market: then the requisite investment required further accelerates the already buoyant market. Hence a lot of unrelated decisions are forced to march in step. Thus the Kahn–Keynes multiplication of expansive and contractive demand furnishes a crucial missing link for Schumpeter's innovative theory of technological evolution.

The central motive force is the seminal conception of a 'swarm' of innovations. There is considerable agreement that the archetypical innovation begins very weakly; then gradually proves its worth, becomes better known, along with improved design and adaptation to diverse uses; finally it decelerates gradually as it is completely integrated into the economy. Thus it tends to have a quadratic trajectory, happily represented by the logistic.

Structural change and growth is introduced by a logistic 'swarm', k, of 50 years' duration. Schumpeter perceived the basic nature of the problem (cyclical growth) but did not succeed in arriving at a satisfactory formulation, having simply assumed full employment with the higher productivity output. To implement structural change, there must first be investment; the investment increases demand and output, including demand for labour. An increase in employment accelerates demand and output in two ways: more employment increases demand and it also increases the real wage, thus generating a twofold increase in demand. The investment is undertaken to lower cost, primarily labour cost. Innovation acts in two opposed directions on employment: by raising output it increases employment but by technological change it lowers employment per unit of output (higher productivity). The rate of decrease of labour input (increase of productivity) is taken to be proportional to the rate of increase of innovational capacity, $\dot{k}/k$. The real wage (average earnings) is assumed to increase faster than productivity when $v > v^*$ and to increase less rapidly when $v < v^*$ (v^* being taken as 0.90, i.e. 10% unemployment). When $v > (v^* + c)$, with $c = 0.048$, i.e. 5% unemployment, the dynamical control parameter, z, increasingly decelerates v, and with $v < 0.952$ it increasingly accelerates v. Thus the pair v and u are dynamically unstable but in economics this constitutes no problem since full employment is an impenetrable barrier, implemented here as approaching full employment for $v > (v^* + c)$. There is first the existing limit of capacity, but that can, given time, be overcome. What cannot be extended is the maximum available labour force. As v goes above $v^* + c$, it becomes rapidly more difficult to recruit labour and hence increase output.

To understand the nature of the system dynamic it is helpful to see it as a simplified variant of the Lotka–Volterra predator–prey model, with wages as the predator and profits the prey. Setting $v^* = 0$, thus reckoning in deviations from equilibrium, the system is a linearized version of Lotka–Volterra with zero constants. Also it is independent of scale and hence not limited to a stationary average level. The central dynamic is given by $\dot{v} = -du$ and $\dot{u} = +hv$, d and h positive. To the first equation must be added a term $fv, f > 0$, which makes the model structurally stable and dynamically unstable, thus answering Kolmogorov's criticism of Volterra's structurally unstable formulation. The term can represent the accelerator or any other aspect of the economy deriving from the simple fact that high and expanding demand leads to further increase in demand and output.

Given an unstable cycle, the usual solution, following Poincaré, is to assume upper and lower non-linearities, yielding global stability and at least one closed limit cycle, a single equilibrium motion. Just as Poincaré generalized the concept of equilibrium from a point to a closed curve, Lorenz generalized the closed curve to a closed region, in which an astonishing variety of aperiodic motions can occur in a seemingly erratic fashion. I shall use a variant of the Rössler Band, as the chaotic attractor most comprehensible and applicable to economics. Instead of defining an upper and lower bound to a variable, one posits a control parameter which provides a growing downward pressure beyond a given high (positive) value and a growing upward one for low (negative) values. The control parameter is specified by

$$\dot{z} = b + gz(v - c).$$

This effectively stabilizes the system globally and leaves it free to perform wildly erratic motion locally around the zero equilibrium. This is illustrated in Fig. 7.1 for a wide variety of initial values resulting in a variety of trajectories which establish an evident boundary. Trajectories from outside all eventually enter the closed region and no trajectory exits from the bounded region. This is such a breath-taking generalization of the notion of a stable equilibrium as to negate partially the conception of equilibrium, which originally suggested absence of change. The concept of a system stable to an equilibrium fixed point has thus been generalized from a point to a closed curve, and then to a bounded region. The closed curve defined a limit cycle, the sole way in which to explain the existence of a cycle. All models from Rayleigh and van der Pol onwards have posited upper and lower values to convert local instability into global stability. They did this by having two non-linearities, commonly in the form of a cubic. Many years ago I realized that one non-linearity (full employment) would also provide a limit cycle, thus demonstrating that two non-linearities were sufficient but not necessary. It is interesting that Rössler found that he needed only one non-linearity to tame the vastly more erratic gyrations of chaotic motion.†

Within the closed region endogenously deterministic variables can move in a seemingly arbitrary and unpredictable way. It is for this reason that this recent discovery is so relevant to economics. If one takes the time-series generated by such a model and tries to determine the mathematical structure which generated it, one would find that to get a correct estimate of the system there is as yet no technique available, and there may never be. If, however, one knows from other sources the model then one can extract its contribution, leaving the irregularity due to random exogenous shocks in the statistic. It is significant that the discovery of chaos was made in application to forecasting weather, a system highly deterministic but with a bad record for prediction, which should be some consolation to economists for their record of prediction.

In my view the implications of chaos for economics are serious. Economic statistics are pervasively irregular: this has always been ascribed to exogenous shocks. In the behaviour exhibited in Fig. 7.1 there are no exogenous elements; the erratic element is entirely endogenous. Therefore in the future it is necessary to adopt the hypothesis that there are two distinct sources of irregularity in economic statistics, the exogenous and the endogenous. This seems to me to require a reformulation of some econometric procedures.

---

† I would like to acknowledge the crucial importance of the assistance I received from Professor Rössler in my amateurish efforts to understand chaos.

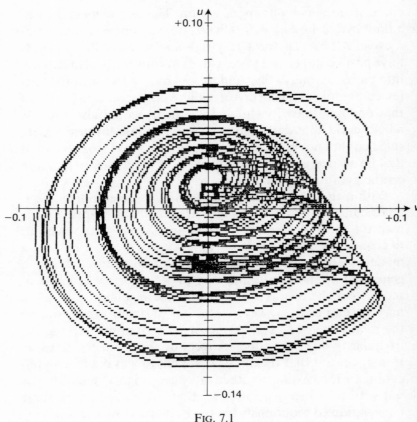

FIG. 7.1

These three differential equations, with only one non-linearity, determine the behaviour of the economy in deviations from the exogenous effects, and, in particular, from the effects of innovatory technical progress. To this basic transmission mechanism, which is independent of scale, must be added the accumulation of innovative capacity as the mechanism of structural change and growth. Also there must be added an equation for the determination of output, as it is affected by innovative investment and as it in turn affects employment, v, and as that in turn affects the share of labour, with its vital effect on consumption demand.

With a constant labour force,

$$v = L/N,$$

so that

$$\dot{v}/v = \dot{L}/L = \dot{q}/q + \dot{a}_L/a_L,$$

with the result that

$$\dot{q}/q = \dot{v}/v - \dot{a}_L/a_L.$$

The rate of change of productivity is taken to be proportional to the accumulation of innovative capacity, thus $m\dot{k}/k = -\dot{a}_L/a_L$. Given historically is a 50-year logistic of innovation:

$$\dot{k} = jk(1 - sk)$$

with the investment reaching a peak in 25 years and approaching zero as $k \to 1/s$. The complete model then becomes:

$$\dot{v} = -du + fv - ez$$
$$\dot{u} = hv$$
$$\dot{z} = b + gz(v - c)$$
$$\dot{q}/q = (-du + fv - ez)/(v + v^*) + mj(1 - sk)$$
$$\dot{k} = jk(1 - sk).$$

The model is not very robust for single parametric changes, but is quite robust for economically plausible related parameter variations. A wide variety of behaviours, generically similar, can be generated. The following set of parameters is chosen solely as a plausible illustration: $d = 0.50$, $e = 0.80$, $h = 0.50$, $f = 0.15$, $b = 0.005$, $g = 85.0$, $c = 0.048$, $m = 0.16$, $j = 0.17$, $s = 0.14$, $v^* = 0.90$. Initial conditions are: 0.020, 0.030, 0, 5.0,

0.045, with $0 < t < 50$ years. The resulting trajectory of output over 50 years is shown in Fig. 7.2. In my view it represents the joint insights of Schumpeter and Keynes and represents what the former wanted to say but could not formulate. He maintained that at the end of a Kondratiev there would necessarily be a higher output since there would always be full employment at a higher productivity. After Keynes one can no longer accept that; there can be those unemployed by labour-saving technology, along with a possibly lower or only moderately higher output and income.

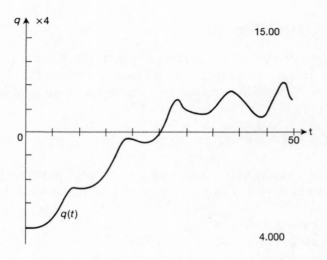

FIG. 7.2

Here there is no assumption of full employment and yet there is a guarantee of a higher average level of output. The problem is posed by the gradual end of the logistic with investment approaching zero. Without further assumptions, barring arbitrary exogenous investment or public policy (unemployment payments are ignored in the interest of sharpness of the issue), there will be a decline of output to its original level or even lower because of labour-saving. Here the problem is solved by the realistic, un-Marxian assumption that the competition of producers for given supplies of labour raises the real wage. With the Keynesian approach, i.e. that demand determines output, the investment demand raises output, this increases employment and demand further, and in a twofold way, since it also means higher demand per employee. This securely confirms that the lowered demand for labour will be more than compensated for.

The Kondratiev peak at a half century is explained by the unimodal quadratic investment function. The rising wave substantially erases the first cycle; then as the Kondratiev levels off towards the end, the cycle slowly re-emerges, and gradually completely takes over as the innovations cease. This downside of the Kondratiev investment is crucial: employment is being cut, though at a decreasing rate, investment demand is rapidly declining. What has happened is that the equilibrium level of output has been shifted upward through the operation of the labour market. In the middle the employment ratio is biased above its equilibrium; this raises demand and the level of wages. Then as the upward motion decelerates, the equilibrium level of employment remains at 10% unemployment of a constant labour force with higher productivity and corresponding real wage and hence demand. Thus Schumpeter's 'vision' is confirmed: there are only slightly less than five cycles along with one large one; the real wage, or average earnings, has risen as has output; structural change has been accomplished but in a cyclical form. What he did not say, but no doubt would have agreed with, is that each cycle is individual, is different from all the others, and not only because of innovations. He confidently asserted that once one Kondratiev was completed, a new wave would get under way. If the new wave is the same as the old, the result is shown in Fig. 7.3. It evidently recapitulates the process on a larger scale, but close inspection will show that each cycle is different in the two periods. In this form the Kondratiev brings history into economics and it is an attractive feature of the model that each successive long wave can be as different as time and technology

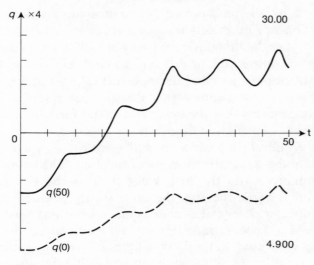

Fɪɢ. 7.3

choose to make it. By varying the parameter, s, one makes the innovative capacity greater or smaller; by varying j correspondingly one determines the rate of growth of productivity; by altering m one makes the economic effect (through the capacity/output ratio) greater or smaller. A related fact is that, for historical reasons, some economies grow fast, for example Japan, and others grow slowly, for example Great Britain. By altering the parameter j from 0.17 to 0.22 and s from 0.14 to 0.10 one gets the high-growth, Japanese case and by altering j to 0.15 and s to 0.25, one gets the low-growth case, with a middle result given by the previous values unaltered, in Fig. 7.4.

The degree of irregularity in these examples is small, whereas the virtue of the model is that it can produce any degree of irregularity. The regularity is more evident in the time-series and phase portrait of the variables u and v, shown in Fig. 7.5. By the modest change of initial values of v and u from 0.02 and 0.03 to 0 and 0.005 and f to 0.18 one arrives at the strikingly irregular resulting time-series and phase portrait

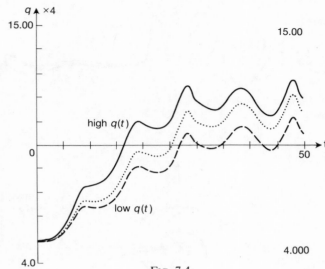

FIG. 7.4

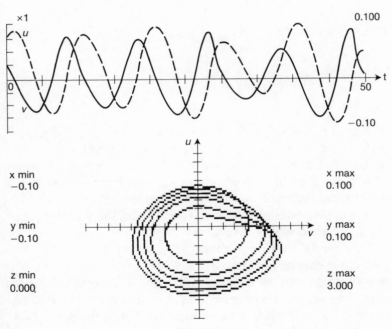

FIG. 7.5

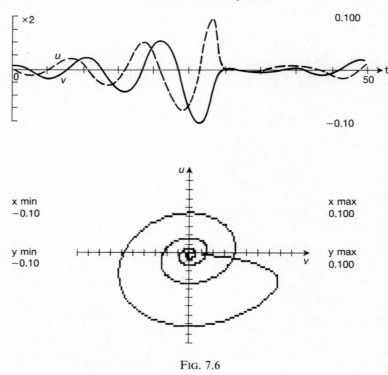

FIG. 7.6

of Fig. 7.6. These produce the rather more irregular output growths exhibited in Fig. 7.7. These two examples appear to me to exhibit the generic character of the non-repeating figures which one sees in economic time-series. The very short-run wiggles can happily be ascribed to random shocks, whereas these irregularities represent a type of dynamic functioning of a simplified economic structure.

This model can be elaborated in varying ways but one particular change is especially important. The logistic innovative function should be bilaterally, not unilaterally, coupled with output, i.e. investment heavily influences demand and output, but is in turn made subject to their influence. Therefore instead of assuming a constant growth rate, j, of new

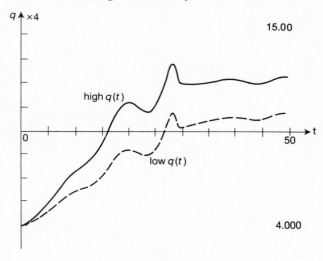

FIG. 7.7

innovative capacity, we may assume a linear dependence on the employment ratio (the state of demand), thus

$$\dot{k} = (j + nv)k(1 - sk),$$

and

$$\dot{q}/q = (-du + fv - ez)/(v + v^*) + m(j + nv)(1 - sk).$$

In this form we have a truly unified, single theory of growth with fluctuation. The degree of irregularity can be indicated by growth rates: they were, peak to peak, roughly 26%, 33%, 6%, 5%, and trough to trough, 19%, 28%, 22%, 23%, or averaging peak and trough, 22.4%, 30.7%, 13.8%, 14.1%. The length of each fluctuation trough to trough (Figs 7.8 and 7.9) was roughly 8 years, 15 years, 7 years, 10 years. Perhaps an even clearer indication of the erratic quality is given by employment $v(t)$ and $u(t)$ in Fig. 7.10.

The foregoing model is a chaotic attractor and is obviously dynamically stable; but is it structurally stable? This is a difficult question and one beyond the scope of this book. To give

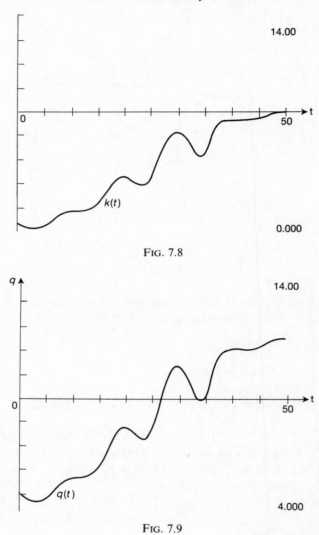

14.00

0.000

FIG. 7.8

14.00

4.000

FIG. 7.9

a simple answer to an unsimple question, I would say it is structurally unstable, which is why it is so interesting. The interested reader is referred to the book by Guckenheimer and Holmes for an extensive treatment of the issue.

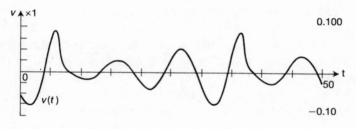

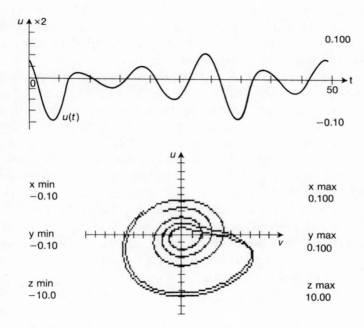

FIG. 7.10

This book, being concerned with chaotic growth through structural change, has followed Schumpeter's un-neoclassical, semi-Marxian vision of growth driven by the search for ever-renewed profit through technical change. However, he inverted Marx by incorporating a rising real wage, with profit, in the stationary state, equal to zero but kept positive by continual, cost-reducing innovations leading to cyclical growth. In spite of some theoretical difficulties, he had a quite realistic insight into how private capitalism works; he was no doubt helped by his disastrous experiences as Finance Minister and private banker.

If I may, in turn, invert Schumpeter, on a basis of no realism whatsoever, imagine a society based on the lust, not for wealth but for power; a society which is as efficient as managerial capitalism and as devoted to technological innovation. Being politically democratic and egalitarian, it would seek to maximize, not growth of output, but the overall well-being of its citizenry. This should be defined and measured by the free choice between work and leisure, but a free choice undistorted by the media of the press, TV, radio etc. In private capitalism these media are directed to increasing output and profit.

In such a utopian community, there could be free individual choice between getting and spending and all the other activities, interests, or pursuits of people. Such a result, being difficult, there would probably have to be a democratically arrived at, common decision setting the number of hours of work per week or per year. In that case, the supply of labour becomes a decision variable, even for a constant population, and must be measured in man-hours not number of workers. Given such a decision, then the problem becomes one of determining output subject to a declining supply of labour, consequent on a rising productivity and a rising average earning. Since $v = a_L q/n(t)$, $\dot{v}/v = \dot{q}/q + \dot{a}_L/a_L - \dot{n}/n$. With productivity growth of $g_a = m\dot{k}/k$, and that of $n(t)$ as g_n, $\dot{q}/q = \dot{v}/v + g_a + g_n$. The economy, though free market, is subject to some sort of perspective control, the aim of which might be monotonic, but cyclical motion; hence with $\ddot{v} = 0$. Research, being heavily subject to state subsidy and controlled with the aim of steady-state growth, would be implemented so as to give exponential rather than logistic behaviour. With a 2% growth rate of productivity, there could be two extreme cases. With no change in the working week, output grows at the rate of productivity, for example from 10 to 27.2 in 50 years with an equal growth

in average earnings. This probably makes little sense for late
capitalism where there is no longer any need for accumulation
of capital, only substitution of newer equipment for the existing. The other extreme would be output constant with
$g_n = -2\%$, so that in 50 years a working week of 40 hours
could be reduced to 14.7 hours. To see the effects and relevance
of such a choice one need only think of the consequences
for the environment and natural resources. By varying the
proportions of these two policies, all combinations in between
these two extremes would be available.

8

Dynamical Control of Economic Waves
by Fiscal Policy

IN previous chapters the control parameter, z, has been treated as simply a device to produce versions of the irregularity characteristic of the evolution of the economy and its time-series. In this chapter I wish to suggest that z can be given concrete embodiment in the form of public net income-generating expenditure. Amongst the many aspects of modern governments, there is the fact that typical budgetary behaviour has a markedly stabilizing effect on economic performance. A large part of government expenditure is substantially invariant to the short-run changes in receipts, in sharp contrast to the behaviour of producers and consumers. Thus public expenditure on armed forces, police, administration, etc. is not limited by tax receipts, and may even vary inversely, for example unemployment and other public assistance.

Compensatory fiscal policy may appropriately be regarded as a type of dynamic control. It aims, or should aim, to diminish the amplitude of fluctuations in economic activity. As a result of somewhat disappointing post-war performance, there is no hope of reducing fluctuations to zero, or even of holding the degree of employment constant. Compensatory fiscal policy has proved to be a somewhat flawed instrument. There are a number of obvious reasons for this. First there is the time-lag involved in collecting and processing the information necessary to know what the state of the economy is. Then there must ensue considerable time to analyse the results and formulate a policy, which may be either legislative or executive or a combination of the two. Then finally, because of the dynamical nature of the generation of income, still more time must elapse before the final consequences of any policy are actually realized. The result is that any policy may be outdated by the time it is effective and it may even be counter-productive. In fact such policies are capable of making the fluctuations worse. Since it is not possible simply to say employment and output should be such and such and proceed to put the economy there, what is required is an effective policy for a gradual approach to any desired position. The Rössler dynamic control parameter, z, represents in simple form just such a procedure.

Given the Rössler Band

$$\dot{v} = -du + fv - ez$$
$$\dot{u} = +hv$$
$$\dot{z} = b + gz(v - c).$$

With z zero, the fixed points for both u and v are zero. For illustration, let $v^* = 0.90$ so that v is in deviations from 10% unemployment, with the level of u unspecified. To represent the normal government practice, let $c = 0.05$, so that when unemployment is less than 5%, public surpluses are increasing, which has the effect of reducing the level of employment progressively. When unemployment is greater than 5%, the government deficit is increasing thus leading to a rising rate of employment. The real wage, or average earning, is increasing faster than productivity whenever v is positive, i.e. $> v^*$, and it is increasing less rapidly than productivity when v is negative.

Taking these as norms for public policy in the absence of a compensatory policy, the behaviour of the employment ratio is shown over a 50-year period in Fig. 8.1, where the parameter values are as follows: d, 0.50, f, 0.15, e, 0.30, h, 0.50, b, 0.01, g, 85.0, c, 0.05, and initial conditions 0, 0.03, 0.02, 5.0, 0.025. Unemployment reaches a maximum of about 12% and a minimum of 1% over the half century. The phase portrait is given in Fig. 8.2. The degree of irregularity is considerable and makes prediction and hence reliable policy-formation nearly impossible. The control variable has a negative influence on

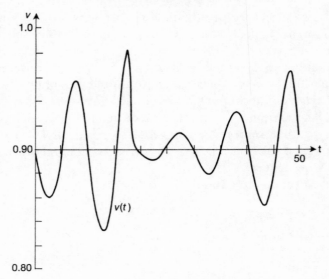

FIG. 8.1

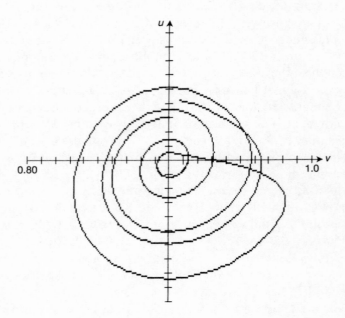

FIG. 8.2

the rate of change of the employment ratio, so that a decline of z signifies an increase of the government deficit and hence an upward motion of output and employment. If the aim is to increase the compensatory element in the budget, the proper thing to do is to lower c, so that corrective action is set in motion above and below, say 7% unemployment. This policy is shown in Fig. 8.3, where the phase portrait is reduced in size

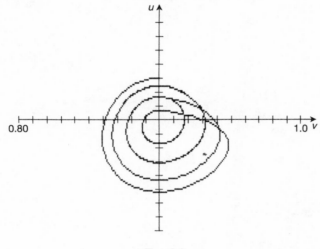

FIG. 8.3

as compared with Fig. 8.2, which means that the amplitude of fluctuation is smaller, with an unchanged economic structure as defined by the other parameters. Emboldened by this modest success, the authorities might try initiating action at $v = 0.02$ (8% unemployment) and strengthen the impact by increasing g from 85.0 to 100.0. They would be well rewarded, as can be seen from a comparison of $v(t)$ and the phase portrait in Fig. 8.4, the latter with $c = 0.02$. Having tasted success, they would surely proceed to try $c = 0.01$. The dramatic result is exhibited in Figs 8.5 and 8.6. The amplitude of the original erratic fluctuations has been ultimately reduced to substantially zero. To exhibit the degree of improvement and its course over time,

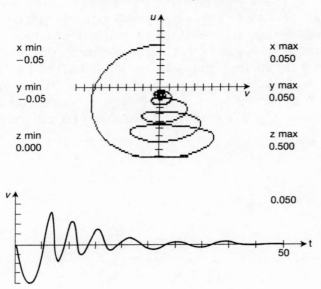

FIG. 8.4

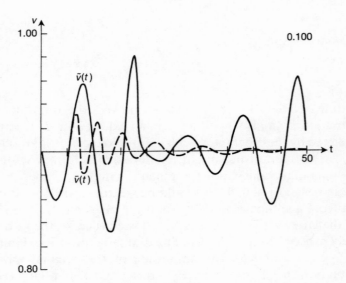

FIG. 8.5

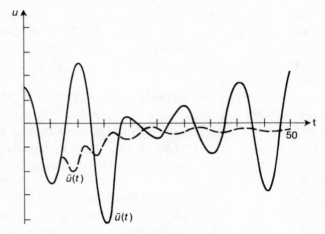

FIG. 8.6

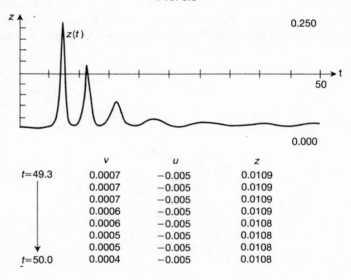

	v	u	z
t=49.3	0.0007	−0.005	0.0109
	0.0007	−0.005	0.0109
	0.0007	−0.005	0.0109
	0.0006	−0.005	0.0109
	0.0006	−0.005	0.0108
	0.0005	−0.005	0.0108
	0.0005	−0.005	0.0108
t=50.0	0.0004	−0.005	0.0108

FIG. 8.7

the series $v(t)$ and $u(t)$ before compensation and $v(t)$ after are presented in Figs 8.5 and 8.6. The requisite magnitudes of compensatory budgeting are shown in Fig. 8.7, where one can see what large and rapid interventions are required in the early stages, but which rapidly diminish thereafter as the fiscal policy becomes increasingly effective in stabilizing output and employment. The last eight iterations during the final year are also given, to demonstrate quantitatively how extraordinarily little fluctuation remains.

The model is then supplemented by including a 50-year logistic of innovation: the results on output are shown in Fig. 8.8. The uncompensated output, $q(t)$, under the impact of the peak investment in the middle of the logistic, has its cycle substantially moderated. But then in the declining phase of the logistic investment, the cycle re-emerges with increased vigour. With the strong compensatory policy ($c = 0.01$), one sees that the sharp interventions in the early period produce added shorter and milder cycles which, successively, smooth the growth path over time. Then as the logistic investment subsides, the fiscal policy has its triumph with sufficient control entirely to erase the three final fluctuations. A second example

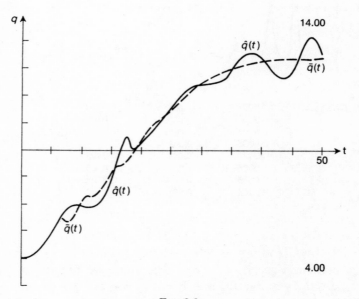

Fig. 8.8

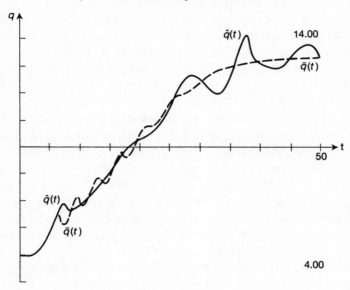

FIG. 8.9

with different initial conditions is given in Fig. 8.9, where qualitatively similar behaviour results.

That such a fiscal policy can be astonishingly successful in smoothing the erratic behaviour of the economy is surely clear. And that some such result is highly desirable is also clear. However, this result has been achieved by stabilizing at 10% unemployment, which is not highly desirable. The foregoing policy is one of 'derivative control' through z. This needs to be accompanied simultaneously by 'proportional control', i.e. policies designed gradually to raise the equilibrium level, v^*, from 0.90 to any desired and feasible level of v, say 0.98. This can be done by successive increases of government or other expenditures by amounts which progressively reduce to zero the difference between actual and desired levels of unemployment (as appears to have happened as a result of the Reagan military budget). Also required is the deployment of monetary policy to avoid the inflationary result of the 1960s, since control of lending is so much more effective in restricting demand than in expanding it. Furthermore there must be a successful policy of forcing or persuading employers and trade unions to forgo raising real wages in consequence of tightness of the labour market, in exchange for high and stable employment along with rising output.

The much more serious problems of implementing such a programme lie in the required magnitude and rapidity of the interventions. However, if it were once agreed that such spectacular improvements might be possible, then it is conceivable that elaborate plans for increased and decreased government expenditure could be worked out and agreed to in advance, so that when the need arose they could be put in motion automatically. It should be admitted that, on the basis of past experience, there is, politically, a very strong asymmetry between deficits and surpluses of government expenditure: surpluses seem to be politically poisonous! The enormous potential of some such flexible, adaptive policy must eventually have an impact, the difficulties notwithstanding. The example elaborated here is quite abstract, excessively simplified, and impractical, but the basic logic of cybernetic control remains central to the nature of fiscal policy.

9

A Fresh Look at Traditional Cycle Models

THE Great Depression of the 1930s prompted the emergence of precise, quantitative business-cycle analysis. The two most influential examples were the Hansen–Samuelson, multiplier-accelerator model and the related Lundberg–Metzler inventory cycle. Hansen, having perceived that the multiplier combined with the accelerator provided the necessary structure, persuaded Samuelson to give it an elegant mathematical formulation. Lundberg, having spelled out producer's behaviour about stocks, left it to Metzler to give mathematical form to the simplest, best-attested trade cycle. Both models, however, were linear and hence required shocks to explain their persistence and irregularity. Later alternative solutions in the form of non-linear dynamics were provided by Kaldor, Hicks, and Goodwin. Such procedures could yield stable, and hence persistent, limit cycles but lacked irregularity. The revival of concern about cycles in the 1970s brought additional cycle models, which, however, were linear and hence needed exogenous shocks to explain their existence. In parallel, the standard econometric methodology assumed linear dynamic structure leaving all irregularity to be attributed to shocks. The discovery, by Lorenz in the 1960s, of 'strange attractors' as explanation of the unpredictability of the weather, led to a number of novel analyses of cycles. They can achieve two fundamental results of prime importance to economics: the motions become bounded and thus globally stable, so that, secondly, the behaviour is aperiodic with a degree of irregularity depending on the parameters.

Given this history, it is illuminating to take a new look, in the light of chaotic dynamics, at the two archetypical cycle models of the 1930s, reformulated non-linearly in such a way as to exhibit persistent irregularity. Such systems give an endogenous, alternative explanation of irregularity, which in no way excludes shocks. In fact, such models introduce shocks in a novel and interesting way. A shock constitutes a new set of initial conditions, and, for non-linear models, initial conditions can produce quite striking alterations of behaviour (bifurcations).

The essence of the Metzler model can be simply formulated in differential form. The rate of change of output, $\dot{q}$, is a proportion, β, of the difference between actual stocks and desired stocks, s^*, which is a proportion, α, of output, thus:

$$\dot{q} = \beta(s^* - s) = (\alpha\beta q - \beta s).$$

The rate of change of stocks is output less demand, which is a proportion, a, of output, thus

$$\dot{s} = q - aq.$$

This system will produce simple harmonic motion with expanding amplitude. If it be expanded to incorporate any one of a number of types of plausible non-linearities, there results a different kind of behaviour. To exemplify this, one can take the Rössler Band, where z may be associated with the perverse behaviour of net public surpluses and deficits. Then

$$\dot{q} = \alpha\beta q - \beta s - ez$$
$$\dot{s} = (1 - a)q$$
$$\dot{z} = \delta + z(q - c).$$

The aim is to produce a cycle of approximately three years; this results from the following suitable parameters: $\alpha = 0.0833$, $\beta = 5.0$, $1 - a = 0.571$, $e = 3.0$, $\delta = 0.10$, $\mu = 2.0$, $c = 2.0$, with initial values of 3.0, 1.0, 0.05. For 15 years the result is shown in Fig. 9.1 with trajectory in state space and time-series of output. This highly erratic behaviour is produced by an undis-

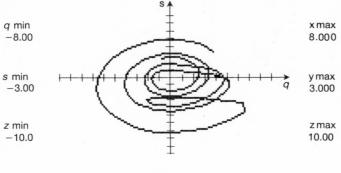

q min		x max
−8.00		8.000
s min		y max
−3.00		3.000
z min		z max
−10.0		10.00

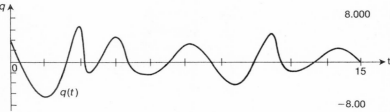

q(t)

8.000

15

−8.00

FIG. 9.1

turbed, entirely endogenous system. Without prior specification of the model, it would be impossible to predict future from past behaviour. If such a system exists, the fact of repeated exogenous shocks poses a difficult problem for econometric analysis. Lacking prior knowledge of the system's parameters, it is difficult to see how one could determine them from a disturbed time-series. As an example, the identical system is shown in Fig. 9.2 with a single initial shock to output of −0.3. The question is: could one derive from this time-series the above parameters?

To represent the Hansen–Samuelson model in simple differential form, I shall assume adaptive dynamic structure. Real investment, $\dot{k}$, is a proportion, γ, of the difference between output and capacity, k, thus

$$\dot{k} = \gamma(q - k).$$

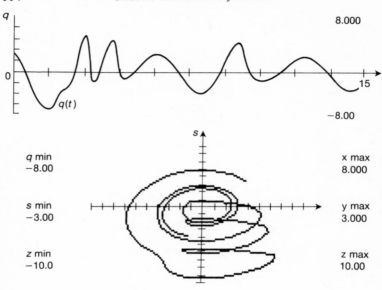

FIG. 9.2

The rate of change of output is a proportion, μ, of the difference between output and demand, where demand is $aq + \kappa \dot{k}$, κ being the capacity/output ratio, hence

$$\dot{q} = \mu(\kappa\gamma - (1-a))q - \mu\kappa\gamma\dot{k}.$$

Adding a single non-linearity as in Rössler, one has

$$\dot{q} = \bar{\mu}(\bar{\kappa}\bar{\gamma} - (1-\bar{a}))q - \bar{\mu}\bar{\kappa}\bar{\gamma}\dot{k} - \bar{e}z$$
$$\dot{k} = \bar{\gamma}q - \bar{\gamma}k$$
$$\dot{z} = \bar{b} + \bar{g}z(q - \bar{s}).$$

The aim is to produce a wave of about 10 years over a 50-year period. In Fig. 9.3 are exhibited the time-series for output and the trajectory in state space, with the following parameters: $\mu\kappa\gamma = 0.9$, $\mu(1-a) = 0.371$, $\gamma = 0.33$, $1-a = 0.475$, $e = 1.0$, $b = 0.1$, $g = 4.0$, $s = 2.0$, and initial conditions $+2$,

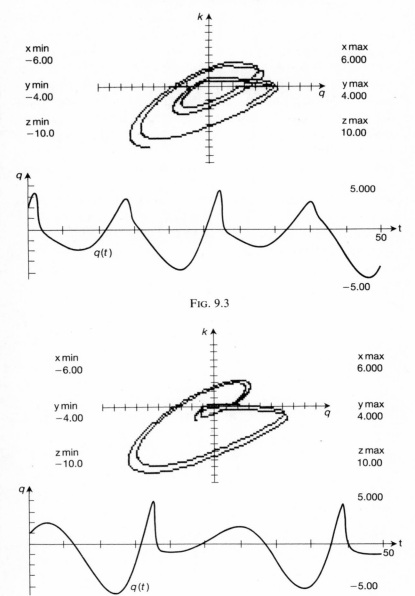

FIG. 9.3

FIG. 9.4

−1, 0.05. Solely by a change of initial conditions, which is equivalent to an exogenous shock, to +1, 0, 0.05, one has the strikingly different behaviour shown in Fig. 9.4.

None of these examples establishes the existence of such behaviour, either in reality or even in the context of such a model. What is demonstrated is that such behaviour is possible and hence cannot be ignored in any econometric study. In analysing dynamic behaviour in the form of time-series, to determine the structural parameters of the economy, it has been customary to explain the irregular residuals as the result of exogenous shocks. These shocks are given arbitrarily and hence can explain anything: therefore, once strange attractors have been discovered, it becomes necessary to consider both types of sources of irregularity in the dynamic behaviour of the modern economy.

10

Chaotic, Aperiodic Behaviour from Forced Oscillators

THE discovery and initial investigation of chaos began not with chaotic attractors but rather with non-autonomous cyclical systems subject to exogenous oscillations. It all began with Poincaré, who in formulating limit cycles also considered fixed saddle-points, which imply such complicated trajectories that he despaired of their ever being understood. His originality only gradually produced two rather different lines of development. The first was initiated by Lord Rayleigh in his analysis of sound: how a single musical note (hence a limit cycle) is produced by forcing a vibrational instrument. Then in the 1920s the development of vacuum-tube oscillators provided van der Pol with a much richer field of investigation. The Rayleigh and van der Pol models are basically similar, differing only in whether stability is controlled by level or rate of change. The van der Pol model is $\ddot{x} + a(x^2 - 1)\dot{x} + bx = 0$. This will obviously produce a limit cycle, being unstable for any $|x| < 1$, and stable for any $|x| > 1$. The quadratic nonlinearity, however, produces special behaviour: for small a the cycle is approximately sinusoidal but for larger and larger a the result approaches a square wave and is appropriately called a relaxation oscillator. Van der Pol, having electronic circuits as his object, was able to make a detailed study of the system, including simulations. Thus he considered the behaviour of his non-linear system under the influence of a variety of exogenous cycles. The system then becomes

$$\dot{x} = y$$
$$\dot{y} = -x - a(x^2 - 1.0)y - b\cos(ct).$$

117

By imposing a suitable exogenous cycle on the van der Pol model one can produce an erratic, chaotic, or aperiodic behaviour, which can become as irregular as one wishes. It is, however, to be distinguished from chaotic attractors: as one can plainly see from simulations, it produces a slowly changing behaviour with a reasonably clear, sequential pattern. This quasi-periodic motion will never, barring special cases, produce exact repetition and hence becomes dense on the torus.

This important feature gives some encouragement to econometricians, faced with the problem of analysing economic time-series. They need to discover the structure of the model which produces the behaviour, so that they can extract the chaotic element and thus isolate the exogenous stochasticity. This aspect may also explain why some mathematicians, long before the discovery of chaos, were making progress in the analysis of forced vibrations. Thus the second line of descent from Poincaré was initiated by Birkhoff in the 1920s, and later pursued by Cartwright, Littlewood, Levinson, and Smale. An analytic approach is much more feasible, given the evident pattern of variation in the oscillations. Without the almost essential help of the computer, Birkhoff was able to predict

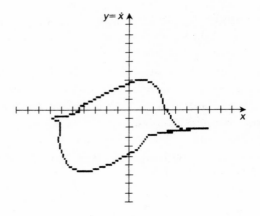

Fig. 10.1

the nature of the phase portrait of such a system. The first simulation of such a trajectory was achieved by R. Shaw, an approximation to which is given in Fig. 10.1. The fact of an evident pattern can be seen clearly in Fig. 10.2 where the phase portrait is simulated over a long stretch of time (5 to 160), with the resulting, slowly varying periodicity exhibited from $t = 55$ to $t = 75$. A second and quite different example of the phase portrait is given in Fig. 10.3 for time 5 to 210.

The time-series corresponding is also included for the period

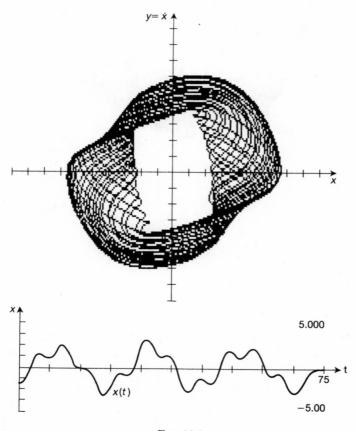

FIG. 10.2

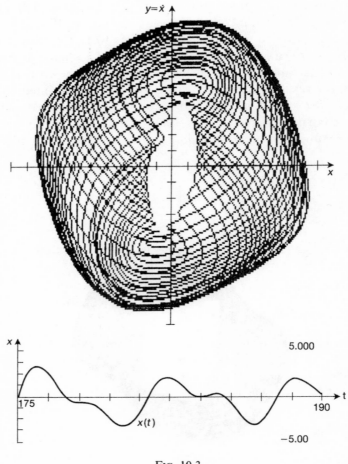

FIG. 10.3

175 to 190. One sees that there is a dominating cyclical pattern regularly produced, but always with a moderate variation.

The characteristics and degree of irregularity depend not only on the ratio of exogenous to endogenous periodicity, but also on the degree of non-linearity in the van der Pol model, as determined by the parameters. If *a* is small, the van der Pol model is rather sinusoidal, so that the impressed wave merely

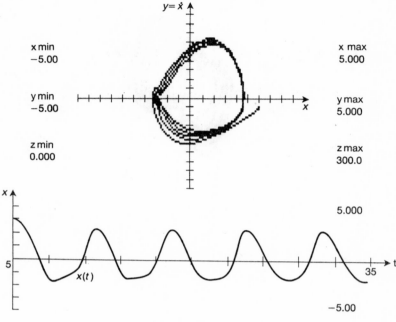

Fɪɢ. 10.4

alters the sinusoid, as is illustrated in Fig. 10.4. Increasing *a* from 0.5 to 1.0 significantly alters the wave shape and increases its variability, as is shown in Fig. 10.5, where the phase portrait is given for time 5 to 95. A further increase of *a* to 2.0 emphasizes the relaxation aspect of the endogenous oscillation. If the parameter *c* is reduced to 0.2, implying a much slower forcing function, the phase portrait is considerably altered and the variation of the wave shape much increased, as illustrated in Fig. 10.6. As is well known, initial conditions can produce much more important changes for non-linear equations than for linear ones. Thus, leaving all parameters unchanged, but altering initial conditions from 3,3,0 to 0,3,0 produces the striking alteration in the behaviour as given in Fig. 10.7.

At this point it becomes appropriate to consider the relevance, if any, that these forced models have to economics.

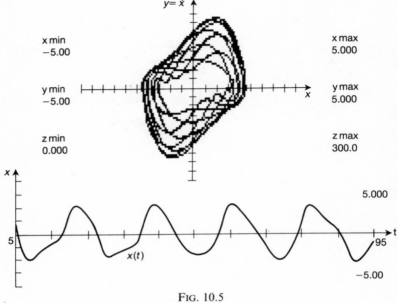

FIG. 10.5

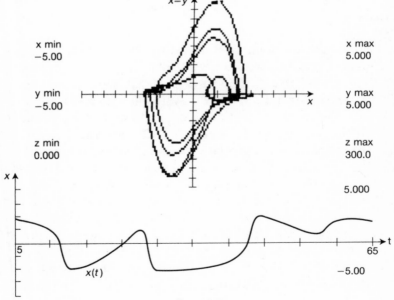

FIG. 10.6

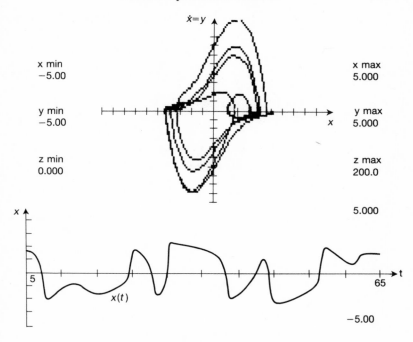

FIG. 10.7

The answer is not difficult to find: the economy consists of a very large number of separate and distinct parts, with the result that these parts are subject to continual exogenous forces. To begin with there are the individual national economies increasingly acted on by the movements of the world economy. Then within the economy there are various markets with dynamics particular to them. There is the annual solar cycle with its influence on various markets, for example the agricultural, the touristic, the fuel, and any number of others. Then there is the pig cycle and the building cycle and, no doubt, numerous others.

As an example, I propose to take the Rössler model of the economy, as presented earlier. Since that model is already a chaotic attractor, it is not exactly obvious what one can expect by impressing on it an exogenous cycle. The equations are as follows:

$$\dot{v} = -u - z$$
$$\dot{u} = +v + bu + c(\cos(w))$$
$$\dot{z} = +b + z(v - a)$$
$$\dot{w} = d.$$

To see an example of the consequences of an impressed wave, one can first simulate the model with no impressed force, i.e. with $c = 0$. Let the other parameters be: $a = 4.0$, $b = 0.2$, $d = 0.8$. In Fig. 10.8 the phase portrait is given for time 25 to 150 and the time-series of v for time 25 to 75. The endogenous model exhibits the doubling of doubled periods, along with a moderately erratic quality, as one can see by a close inspection

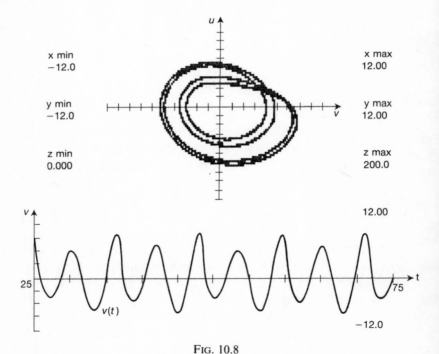

FIG. 10.8

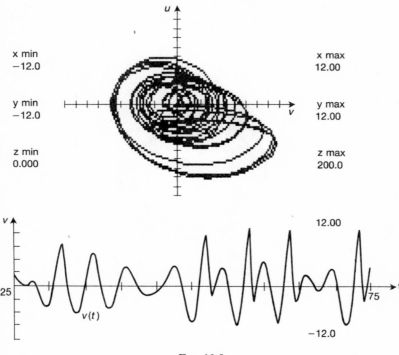

FIG. 10.9

of the time-series of $v(t)$. By altering the value of c to 1.5, one can see the dramatic effect of impressing a force of a different, single periodicity. In Fig. 10.9 is given the effect with the same initial values and the same parameters other than c, over the same two periods. The relatively simple shape of the phase portrait is quite destroyed and time-series of $v(t)$ bears almost no resemblance to the autonomous behaviour. In Fig. 10.10 is given the phase portrait and time-series for the parameter values $a = 3.0$, $b = 0.2$, $c = 0$, and $d = 0.8$, to show the behaviour in the absence of forcing. In order to avoid any transient response to the initial conditions, the phase portrait is simulated for time 25 to 225, with the result of a double period within a narrow band, indicating a very modest irregularity.

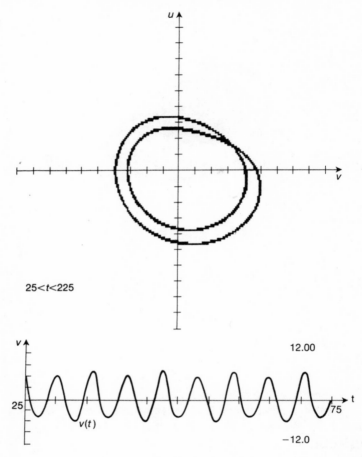

25<t<225

FIG. 10.10

The time-series for $v(t)$ shows the mild erratic characteristic. With no change except introducing the forcing function ($c = 2.0$), the comparable phase portrait and time-series is given in Fig. 10.11. The truly astonishing change in both figures is impressive evidence for the effect of forcing the economic Rössler. The impression given is of a disturbed chaotic attractor or a kind of super-chaos. The system appears never to repeat even over such a long period.

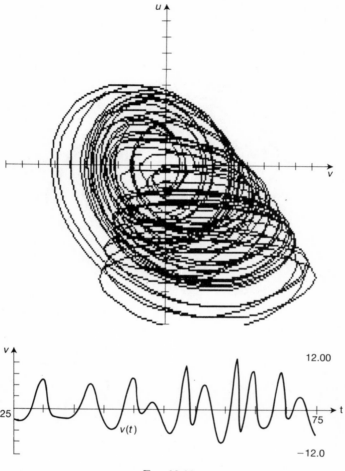

Fig. 10.11

The previous example was one of a very mild irregularity, so that the astonishing increase of chaotic behaviour in Fig. 10.11 might be thought to be simply the result of the particular forcing function. To demonstrate that it is not so, a second more erratic attractor is simulated in Figs 10.12 and 10.13. First there is the autonomous model with no impressed force,

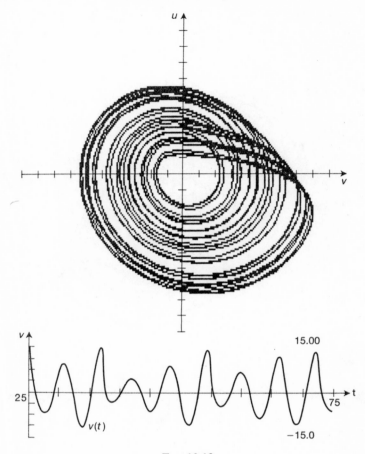

FIG. 10.12

by setting $c = 0$. In Fig. 10.12 the parameters are $a = 6.0$, $b = 0.2$, and $d = 0.8$, with a run of time from 25 to 225; evidently the system has a full range of chaotic behaviour, which is also shown in the time-series of $v(t)$ for 25 to 75. Then in Fig. 10.13, with no change except the introduction of the forcing function by setting $c = 2$, the extraordinary alteration of the phase portrait in both periodicity and amplitude is produced. This is further evidenced by the greatly increased irregularity

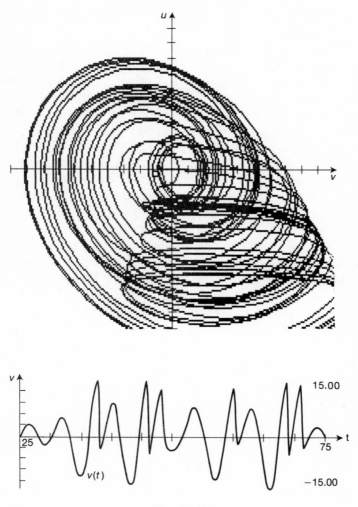

FIG. 10.13

in $v(t)$. Thus, very surprisingly, given a strongly chaotic attractor, the impressing on it of a single, exogenous periodic, results in a striking alteration and increase of the chaotic behaviour of the system. In view of the prevalence of this kind of situation in most economies, this fact is surely of importance in understanding the markedly erratic behaviour of the parts of modern economies.

The modern economy consists of a great variety of separate activities intimately linked, directly or indirectly through markets, with all or most of the other sectors. Thus international trade has grown faster than national trade, and the proliferation and reduction in cost of internal transport has increasingly unified national markets and producers. This means that the various producing units are continually subject to erratic variations in both their supplies and sales. For these reasons the analysis of the consequences of forcing functions (that is, all relevant events beyond the control of the producer) on the behaviour of production is an important addition to our understanding of the extremely complex interdependence of the modern economy. It is precisely the erratic character of economic events that creates the problems and difficulties faced by producers, since their current decisions depend so heavily on an unpredictable future.

Further reading

Abraham, Ralph H., and Shaw, Christopher D. (1981), *Dynamics, the Geometry of Behaviour; pt. 2: Chaos* (Santa Cruz, Aerial Press).

Arnold, Vladimir Igorevich (1978), *Ordinary Differential Equations* (Cambridge, Mass., MIT Press).

—— (1983), *Geometrical Methods in the Theory of Ordinary Differential Equations* (Berlin, Springer).

Devaney, R. L. (1986), *Introduction to Chaotic Dynamical Systems* (Menlo Park, Calif.: Benjamin/Cummings).

Guckenheimer, John, and Holmes, Philip (1983), *Nonlinear Oscillations, Dynamical Systems, and Bifurcations of Vector Fields* (Berlin, Springer).

Hirsch, Morris W., and Smale, Stephen (1974), *Differential Equations and Linear Algebra* (New York, Academic Press).

Holden, A. V. (ed.) (1986), *Chaos* (Manchester, Manchester University Press).

Lorenz, H.-W. (1989), *Nonlinear Dynamical Economics and Chaotic Motion* (Berlin, Springer).

Minorsky, N. (1962), *Nonlinear Oscillations* (Princeton, Van Nostrand).

Smale, Stephen (1980), *The Mathematics of Time; Essays on Dynamical Systems, Economic Processes* (Berlin, Springer).

Thompson, J. M. T., and Stewart, H. B. (1986), *Nonlinear Dynamics and Chaos* (New York, John Wiley).

Index